中生代新疆陆地生态景观复原示意图
Reconstuction of a Mesozoic Xinjiang landscape

国家科学技术学术著作出版基金资助出版

# 中生代新疆

Mesozoic of Xinjiang, China

上海科技教育出版社

## 出版说明

科学技术是第一生产力。21世纪，科学技术和生产力必将发生新的革命性突破。

为贯彻落实"科教兴国"和"科教兴市"战略，上海市科学技术委员会和上海市新闻出版局于2000年设立"上海科技专著出版资金"，资助优秀科技著作在上海出版。

本书出版受"上海科技专著出版资金"资助。

上海科技专著出版资金管理委员会

# Mesozoic of Xinjiang, China

# 中生代新疆

孙革　莫斯布鲁格　孙跃武　阿什拉夫　马丁
著
苗雨雁　王克卓　吴文昊　杨涛　董曼

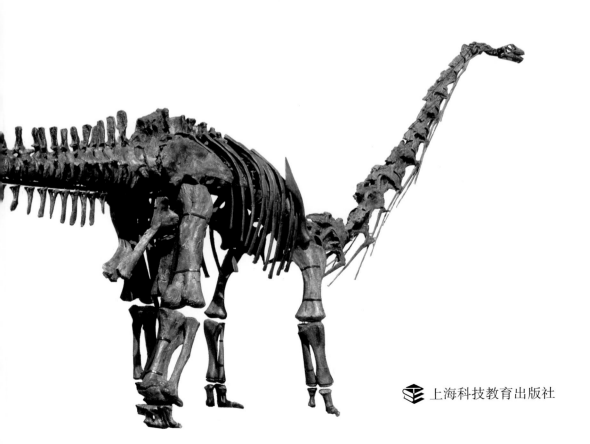

上海科技教育出版社

**图书在版编目（CIP）数据**

中生代新疆/孙革等著. —上海：上海科技教育出版社，
2022.12

ISBN 978-7-5428-7849-6

Ⅰ.①中… Ⅱ.①孙… Ⅲ.①地层古生物学—研究—新
疆—中生代 Ⅳ.①Q911.745

中国版本图书馆CIP数据核字（2022）第203337号


**责任编辑** 伍慧玲
**封面设计** 汤世梁
**版式设计** 李梦雪


ZHONGSHENGDAI XINJIANG
**中生代新疆**

孙革 莫斯布鲁格 孙跃武 阿什拉夫 马丁
苗雨雁 王克卓 吴文昊 杨涛 董曼 著


**出版发行** 上海科技教育出版社有限公司
（上海市闵行区号景路159弄A座8楼 邮政编码201101）

| | | |
|---|---|---|
| 网 址 | www.sste.com www.ewen.co |
| 经 销 | 各地新华书店 |
| 印 刷 | 常熟市华顺印刷有限公司 |
| 开 本 | 720×1000 1/16 |
| 印 张 | 18 |
| 版 次 | 2022年12月第1版 |
| 印 次 | 2022年12月第1次印刷 |
| 书 号 | ISBN 978-7-5428-7849-6/N·1166 |
| 审 图 号 | GS（2022）5475号 |
| 定 价 | 160.00元 |

# 内容提要

新疆位于中国西北边陲，是一个神奇而又令人向往的地方。中生代时期（距今约 2.52 亿～0.66 亿年）这里气候温暖湿润，森林密布，河湖纵横，为恐龙等各类生物的繁衍生息，以及煤和石油等化石能源的形成创造了诸多有利条件。

新疆地层出露完整，化石丰富，为研究我国北方中生代地质与生物演化提供了一个理想场地，也成为"一带一路"沿线及相关国家的地学家们长期关注的地方。20 多年间，古生物学家孙革和莫斯布鲁格率领的中德合作科研队在新疆开展了卓有成效的中生代地层古生物合作研究，取得了一批具有国际意义的重要研究成果，如首次发现迄今最大的侏罗纪恐龙"新疆巨龙"、迄今最大规模的鄯善侏罗纪恐龙足迹群、早三叠世水龙兽、侏罗纪准噶尔兽等哺乳动物新组合、新疆龟新种、郝家沟晚三叠世植物群，以及对新疆中侏罗世白杨河植物群及沙尔湖植物群等开展的深入研究。上述成果对推动新疆中生代生物演化、地层对比研究，以及找寻石油、煤等能源矿产等具有重要科学意义。

本书通过中德合作科研队 20 余年来在新疆共同开展研究及紧密合作的故事，生动介绍了新疆地区的中生代地层古生物研究的最新进展和中德科学家们工作的艰辛，许多故事感人至深；书中还首次推出近 20 余年来新发现的"新疆中生代十大化石明星"，深度介绍了中德两国年轻科学家在新疆的成长历程。总之，本书对推动我国中生代地质古生物研究、科学普及及促进中德科学合作将发挥重要作用。

# 目　录 ────────────────────────────

# CONTENTS

奥勃鲁契夫银杏

*Ginkgo obrutschewi* Seward

# 前　言

　　新疆位于中国西北边陲，面积约 166 万平方千米，占国土总面积近 1/6，是中国地理面积最大的省区，也是"古丝绸之路"的重要通道，以及连接中亚、俄罗斯和蒙古西部的纽带。中生代时期（2.52 亿 ~0.66 亿年前）的新疆囊括准噶尔、塔里木和吐鲁番 – 哈密（简称吐哈）三大盆地，孕育了丰富的石油、煤及天然气等化石能源矿产。中生代的新疆宛如一本地质教科书：站在天山之巅眺望，银白色雪山（多为 3 亿多年前的石炭纪地层）之下，绵延起伏的低山大都由中生代地层组成，色彩斑斓、交叠展布，宛如一幅幅精美的画卷，令人惊赞和留恋。

　　新疆三大盆地中生代生物化石十分丰富，为研究中国新疆和西北乃至中亚地区中生代生物演化与环境变迁提供了珍贵的资料。这些化石生物又为新疆中生代石油、煤炭等的形成提供了宝贵的物质来源。因此，新疆奇特的地质古生物资源，特别是中生代生物化石及其所展现的古环境变化等，早已引起了国际地学界的广泛关注，新疆也成为一个理想的"天然实验室"（图 1）。

　　新疆的地质研究工作历史久远，最早可追溯到俄罗斯地质学家奥勃鲁契夫（Obruchev V. A.）等对新疆

# 第 **1** 章

# 追索奥勃鲁契夫之路

## 1.1 追索奥勃鲁契夫考察之路

### 1.1.1 奥勃鲁契夫的准噶尔野外考察

奥勃鲁契夫院士是俄罗斯最杰出的地质学家之一，1886 年毕业于圣彼得堡矿业学院，1901 年起任托姆斯克工业学院教授，1921~1929 年任莫斯科矿业学院教授，1930 年起任苏联科学院冻土研究所所长，1929 年当选为苏联科学院院士，曾荣获多枚列宁勋章及首枚卡尔宾斯基金质奖章。他长期从事西伯利亚、蒙古及中亚的地质研究，1904~1905 年和 1909 年曾两次经蒙古来中国准噶尔盆地进行野外地质考察，采集了大量古生物化石，并首次绘制了《准噶尔地区地质图》（1∶50 万）（Obruchev，1914，1940a）（图 2）。

1904~1905 年，奥勃鲁契夫在准噶尔盆地西北部的狄阿姆河（Diam River）的阿克扎尔河谷（Ak–djar ravine）等地采集了一批植物化石并考察了这里的地质情况。回到俄国后，他将化石寄给了英国大英博物馆（自然史部）著名古植物学家秀厄德院士（Seward A. C.）鉴定。秀厄德鉴定了 16 属 19 种，并建立了 2 个新种，认为这些植物化石的时代属侏罗纪。此后，秀厄德发表了著名论文《中国准噶尔侏罗纪植物》（Seward，1911），奥勃鲁契夫也将化

**图 2 奥勃鲁契夫及秀厄德有关新疆准噶尔的著作及地质图**

Fig. 2 Monographs and geological map of Obruchev and Seward on Junggar of Xinjiang, China

A. 奥勃鲁契夫院士； B. 20 世纪初准噶尔白杨河谷地景观（1904~1905）； C. 奥勃鲁契夫绘制的准噶尔地质图，示化石产地"阿克扎尔"及"狄阿姆河"；D. 奥勃鲁契夫著作( Obruchev，1940a )；E. 秀厄德院士；F. 秀厄德院士的专著《中国准噶尔侏罗纪植物》（1911）

A. Prof. Obruchev; B. Baiyang River valley, 1904–1905; C. Geological map showing the fossil site and river name: Ak–djar and Diam River; D. the work of Obruchev, 1940; E, F. Prof. Seward and his *Jurassic Plants from Chinese Dzungaria*, 1911

石鉴定结果用于他的《准噶尔地区地理与地质》等论著中（Obruchev，1914，
1940b）（图 2，D）。这两位专家都指出，化石产于狄阿姆河的阿克扎尔河谷。
秀厄德在他的上述论文的化石鉴定中还将"狄阿姆"作为一个新种的种本名，
即"狄阿姆拉法尔蕨"（*Raphaelia diamensis* Seward）。后来，在奥勃鲁契夫
绘制的《准噶尔地区地质图》（1∶50 万）中也明确标注了阿克扎尔河谷和狄
阿姆河所在地（图 2，C）。可惜，"狄阿姆"和"阿克扎尔"这两个地名在
中国的地图上找不到，奥勃鲁契夫的地质图也难以查找。半个多世纪过去了，
"狄阿姆"和"阿克扎尔"究竟在哪里，既无人知晓，也无人仔细考究。这
样一来，中国地质古生物工作者的心中长期留下了一个"狄阿姆之谜"。

## 1.1.2 解开"狄阿姆之谜"

　　1988~1989 年，孙革被国家派往英国大英博物馆（自然史部）留学，在
古植物学家秀厄德院士当年的办公室（秀厄德图书室）工作一个月。作为古
植物学工作者，孙革熟知秀厄德对中国准噶尔侏罗纪植物研究的贡献，以往
也经常使用"狄阿姆拉法尔蕨"的学名用于化石鉴定，现在能亲自在秀厄德
当年的办公室工作，除深感荣幸外，对秀厄德有关新疆准噶尔侏罗纪植物化
石的研究更增加了兴趣。孙革在大英博物馆（自然史部）查阅馆藏标本得知，
秀厄德研究完这批准噶尔化石后，标本早已全部返还给奥勃鲁契夫，这些化
石后来一直存放在圣彼得堡的全苏地质博物馆（VSEGEI）；有关"狄阿姆""阿
克扎尔"的位置更无从查找。这一困惑和遗憾一直深存在孙革的脑际，他决
心在回国后想方设法找到这些化石产地，这不仅对进一步开展新疆准噶尔侏
罗纪植物及地层研究大有裨益，也将有助于早日解开这个令人困扰的"狄阿
姆之谜"。

　　1989 年，孙革结束了在英国的进修、返回南京后，恰获老朋友、著名地
质学家沈远超教授之邀，赴新疆野外考察。沈教授是中国科学院地质所研究员、
新疆 305 项目首席科学家，当时正在新疆准噶尔盆地哈图地区主持找寻金矿
的工作。有了这一难得的机会，孙革从南京飞抵乌鲁木齐，之后去奇台请新

疆第一区调大队年轻的古植物学专家李洁帮忙。李洁早年曾在中科院南京古
生物研究所进修,与孙革以师生相称。借助沈教授提供的一辆北京牌吉普车,
孙革和李洁从石油之城克拉玛依出发,开始了追寻当年奥勃鲁契夫院士准噶
尔地质之旅的路程(图3,A)。

**图3 找寻奥勃鲁契夫的白杨河化石点**

Fig. 3 Tracing fossil site Baiyanghe (Diam River) worked by Obruchev

A. 找寻奥勃鲁契夫化石点的行程路线(1989);B. 和布克赛尔;C、D. 白杨河(狄阿姆河)
河谷

A. The route for tracing Obruchev's expedition in Junggar; B. Hoboksar; C, D. Baiyang(Diam)
River valley

孙革等最先抵达的是新疆北部的一个煤炭小镇——和什托洛盖镇，在那里采集了一批化石，了解了一些情况，但无人知晓"狄阿姆"等地名。两天后到达了准噶尔最北部的边境县城——和布克赛尔蒙古自治县（和丰县）（图 3，B），在这里，向在当地政府工作的哈萨克族工作人员了解到，"狄阿姆"（Diam）是哈萨克语"白杨"的意思，在距和丰县大约 200 千米的托里县铁厂沟附近有一个叫"白杨河"的地方。这个消息让孙革等喜出望外，因为这与孙革和李洁事前的分析、他们手中的准噶尔西北部的地质图和地形图上标注的位置十分吻合！于是，他们顾不上欣赏群山怀抱下一望无际的大草原的美丽风光，第二天一大早便离开和丰县，驱车前往托里县铁厂沟，去找寻梦寐以求的"白杨河"。

## 1.1.3　托里白杨河化石产地的发现

铁厂沟是一个小型煤矿所在地，位于和丰县西南约 200 千米，但两地间有横亘东西的谢米斯台山（Semistai Mt.）阻隔。从和丰县到铁厂沟要穿过一个大山坳口，那里曾有一座稀有金属矿山，现已废弃。从和丰县到山坳口之间的公路已经老旧，很少有人使用，特别是过了山坳口之后，就是一望无际的戈壁滩了，几乎无路可行。虽然这些困难对多年从事区域地质调查工作的孙革和助手李洁而言并不在话下，但由于道路难行，区区约 200 千米的路竟整整花费了近一天的时间。他们抵达大山坳口废弃矿山时已是下午。下山后，眼前迎来的是茫茫戈壁滩，已经没有公路了，但向远处望去，能隐隐约约看到有煤矿小镇（铁厂沟）的影子。孙革等更加强了信心，远处的目标也与地图的指向一致，只是这余下六七十千米的路程全是戈壁滩上的"搓衣板"路，车上的颠簸成了一天中最难熬的时刻。不管怎样，胜利最终属于勇敢者。天快黑时，孙革一行终于抵达了铁厂沟煤矿，住进了小煤矿的招待所。

第二天一早，孙革一行三人按照地图指出的方向，由铁厂沟一路向东，顺着戈壁滩上模糊的车辙印或羊群走过的足印，驱车一个多小时，终于见到了一片南北展向的台地。到了台地的顶处，俯瞰谷底，一片茂密的杨树林宛

　　如一条宽阔的绿带映入眼帘，一条蜿蜒的小河自北向南从林中穿过，在密林和小河的东面是被河谷切割露出的完美的含煤地层剖面。梦寐已久的狄阿姆河——白杨河终于找到了（图3，D；图4）！这个神秘的地层剖面和产煤地

**图 4　白杨河剖面（阿克扎尔剖面）**

Fig. 4　Baiyanghe section (Ak-djar section)

A. 白杨河谷剖面南段；B. 河谷东侧的中侏罗世含煤地层；C. 当地哈萨克牧民；D. 孙革；E. 李洁；F. 中德科研队考察白杨河剖面，右起：孙革、迪尔切及莫斯布鲁格等（2001.9）；

A. South part of the Baiyanghe section；B. Middle Jurassic coal-bearing strata；C. local Kazakh herdsman；D. Sun G.；E. Li J.；F. Sino-German research team working in Baiyanghe, from right: Sun G., Dilcher D. L., Mosbrugger V., et al. （2001.9）

点就是隐蔽在奥勃鲁契夫所称的阿克扎尔河谷里，与奥勃鲁契夫当年绘制的
地质图的标记完全一致！从后来与当地哈萨克牧民交谈中得知，原来"阿克
扎尔"的"阿克"（Ak）是"白"的意思，"扎尔"（djar）意为"树"，"阿
克扎尔"就是"白树林"的意思。可能也因为这里长满了白杨树，河名由此
成了狄阿姆河，也就是今天的白杨河（图 4，A、B）。

　　走进白杨河谷，眼前便是多处废弃的土墙。一位放羊的哈萨克老人说，
这里曾有挖煤人和牧民居住。这位看上去有六七十岁的老人还介绍说，他在
这里生活得很安逸，他有 100 多只羊，雇了一位年轻人放养，政府的政策好，
卖羊的收入很可观，因此，他可以天天喝酒，不愁吃也不愁穿，可以在这里
安度晚年（图 4，C）。

　　白杨河谷剖面（即阿克扎尔剖面）十分壮观，南北长近 3 千米，侏罗纪
含煤地层主要由灰色的砂岩及粉砂岩组成，至少含 3~4 个厚煤层。这里的植
物化石众多，几乎俯首可见。完美的地质剖面和奇妙的自然景观浑然一体，
格外令人震撼。受吉普车借用时间所限，孙革等在这里仅工作了两天，却采
得 4 箱百余块保存完好的植物化石，并对剖面做了初步勘查，可谓收获满满。
至此，这个"狄阿姆之谜"终于被解开，孙革等也从此与白杨河和新疆结下
了不解之缘。

# 1.2　波恩及图宾根之约

## 1.2.1　波恩及图宾根之约

　　中国的孙革与德国的莫斯布鲁格（Mosbrugger V.）的友谊从 1986 年开始。
1986 年，在德国波恩大学古生物研究所工作的莫斯布鲁格首次来华，受导师、
著名古植物学家斯崴瑟（Schweitzer H.-J.）教授的委托，赴山西保德考察二叠
纪地层及植物化石。因 20 世纪 20 年代瑞典著名古植物学家哈勒（Halle T.）
来华时在山西保德采集的植物化石的研究工作没做完，瑞典自然史博物馆（位
于斯德哥尔摩）委托斯崴瑟教授继续完成此项工作。莫斯布鲁格当时三十出头，

**图 5 莫斯布鲁格的首次中国之旅**

Fig. 5 The first visit of Mosbrugger in China

A. 中国科学院南京地质古生物所；B. 李星学院士（左）与斯崴瑟教授夫妇在北京（1980）；
C. 莫斯布鲁格（左）与李院士在秦皇岛（2000）；D–H. 莫斯布鲁格（G左）在蔡重阳教授（G
中）陪同下于山西保德扒楼沟煤矿野外考察（1986），E 为考察中新发现的根座化石

A. NIGPAS, Nanjing; B. Academician Li X. X.（left）with Prof. Schweitzer and his wife in Beijing
（1980）；C. Mosbrugger V. with Li X. X. in Qinhuangdao, China（2000）；D–H. Field work
of Mosbrugger（G–left）guided by Prof. Cai C. Y.（G–mid）in Balougou Coal–Mine of Baode,
Shanxi（1986）: E showing the plant fossil *Stigmari* which was newly found

风华正茂，满怀热忱和期盼地踏上了来中国南京的旅程。他此次来华是应中国科学院南京地质古生物研究所（简称南古所，NIGPAS）著名古植物学家李星学院士的邀请。李院士与斯崴瑟教授早就熟悉，曾在 1980 年一同参与在北京举行的青藏高原国际会议。后来，李院士将自己的学生蔡重阳送到德国斯崴瑟教授的实验室，以"洪堡学者"身份进修，蔡教授也是莫斯布鲁格博士的老朋友。为迎接这位年轻的德国古植物学专家，刚获博士学位不久的孙革先行赴山西保德为莫斯布鲁格的野外之行"打前站"。孙革又是蔡重阳研究员的师弟，因此很快也成了莫斯布鲁格的好朋友。莫斯布鲁格在南京期间不仅查看和研究了南古所保存的有关植物的化石，还做了精彩的学术报告，受到大家的欢迎。其间，莫斯布鲁格与孙革也多次交流，还应邀到孙革家中做客，彼此建立起深厚的友谊（图 5）。

1987 年，孙革获首批"中英友好奖学金"并于 1988 年初赴英国大英博物馆（自然史部）进修。同年 9 月，孙革应斯崴瑟教授邀请，前往德国波恩大学地质古生物研究所访问，接待工作由莫斯布鲁格博士负责。两个老朋友久别后在波恩重逢，格外高兴。受斯崴瑟教授委托、参与接待的还有年轻的孢粉学专家阿什拉夫（Ashraf A. R.）博士。阿什拉夫于 20 世纪 70 年代末曾随该所时任所长埃尔本（Erben A.）教授率领的德国专家组来中国广东南雄进行 K-T 界线合作研究，对中国和中国同行充满深厚的感情。为讨论未来中德双方的合作，莫斯布鲁格和孙革先是在实验室共同分析研究了孙革带来的中国霍林河煤田的植物化石，又请阿什拉夫做了孢粉分析。大家都感到，中国的化石保存较为理想。于是，双方约定：未来的中德合作就以中国霍林河含煤地层及其植物化石研究为题。很快，双方共同起草了一份联合申请书，由孙革在中国代表中德双方申请德国大众汽车公司（Volkswagen Co.）资助的科研项目（图 6）。但不巧的是，由于该公司在华资助项目的范围缩减，项目申请未能成功。后来，由于孙革忙于其他科研项目，此项"波恩之约"被搁置下来。

1990 年，莫斯布鲁格应聘德国图宾根大学地质古生物学院院长，开始了

**图 6 德国波恩之约（1988）**

Fig. 6　Appointment in Bonn, Germany in 1988

A、G. 波恩大学地质古生物研究所；B. 英国大英博物馆（自然史部，1988）；C. 孙革在秀厄德图书室工作；D. 斯崴瑟教授设家宴邀请孙革（1988）；E、F. 莫斯布鲁格与孙革在波恩（1988）；H. 研究所的地质古生物博物馆

A，G. Institute of Geology and Paleontology, Univ. Bonn; B. British Museum（NH）in London（1988）; C. Sun G. working in Seward Library; D. Prof. Schweizer inviting Sun G. to family dinner（1988）; E, F. Mosbrugger V. and Sun G. in Bonn （1988）; H. Goldfuβ-Museum of Inst. Geol. Paleont., Univ. Bonn

**图 7　图宾根之约**

Fig. 7　Agreement in Tuebingen, Germany

A. 图宾根大学城堡；B. 地质古生物学院；C. 孙革与阿什拉夫及青年学者麦什和马茨克博士
讨论新疆研究工作；D. 在阿什拉夫家做客，右起：阿什拉夫夫妇，菲尔塞克博士（图宾根
大学东方文化中心主任），莫斯布鲁格，孙革；E、F. 图宾根大学地质古生物博物馆

A. The former castle of Tuebingen City now occupied by Univ. Tiebingen; B. Institute of Geosc.
Paleont., UT; C. Discussion on research work in Xinjiang by Sun G., Ashraf A. R. and young men
Drs. Maisch M. and Matzke; D. Visiting at Ashraf' home, from right: 1, 2. Ashraf A. R. and his wife, 3.
Dr. Filseck K.（Head of Asian Culture Center of UT）, 4. Mosbrugger V., 5. Sun G; E, F. Geol.
Paleont. Museum of the Institute, UT

他科研及教学工作的新天地。1995年，孙革开始担任中科院南古所的领导。双方的合作有了新的契机。1996年6月，孙革来英国曼彻斯特大学完成中科院资助的高访项目，当时正值图宾根大学将举行地质系成立百年庆典。应莫斯布鲁格邀请，孙革再次踏上德国大地，开始了他首次图宾根之旅。庆典活动在图宾根大学的古城堡大厅举行，国际专家云集，格外成功。此次久别重逢，孙革再次与莫斯布鲁格探讨了双方的合作，特别是介绍了他回国后在新疆追索俄罗斯奥勃鲁契夫院士"准噶尔野外考察之谜"的故事。经过同老朋友阿什拉夫磋商，双方约定：以中国新疆准噶尔盆地中生代地层与古生物的合作研究作为中德合作的新项目（图7）。

此时的孙革和莫斯布鲁格在各自的科研岗位上均已取得较多成绩，这为开展中德合作创造了更多的有利条件。恰好，当时正值中国科学院（CAS）与德国马普学会（Max-Planck Soc.）合作项目积极开展，以孙革和莫斯布鲁格牵头的中德合作"新疆准噶尔盆地中生代地层古生物研究"项目的申请，很快获得中德马普合作项目的批准。由此，自1997年起，双方"波恩之约"和"图宾根之约"终于如愿以偿。

## 1.2.2 中德联合新疆科研队的诞生

1997~1999年，在中德马普合作项目的支持下，由孙革和莫斯布鲁格领导的中德合作"新疆准噶尔盆地中生代地层古生物研究"项目顺利开展起来。两国专家首先重点研究了乌鲁木齐附近的郝家沟剖面的晚三叠世地层古生物，取得了可喜成果。与此同时，对整个准噶尔盆地中生代地层（包括奇台将军庙地区、大龙口P-T界线剖面、石河子地区玛纳斯河剖面及红沟剖面，以及克拉玛依西北的白杨河剖面等）进行了初步的野外考察，工作取得重要进展。中德合作新疆项目工作一开始，就得到新疆第一区调大队的大力支持，包括提供相关地质资料、野外车辆及许多接待工作等。

为进一步加强中德双方在中国新疆地学研究的合作，2000年12月1日，由新疆第一区调大队、中科院南古所和德国图宾根大学地质古生物学院共同成立了"中德合作新疆地质工作站"（Sino-German Co-Working Station for Geosciences in Xinjiang，SGWSGX）。左学义（时任新疆第一区调大队队长）、孙革和莫斯布鲁格共同担任主任，成立了由 19 人组成的管委会，工作站办公地点设在新疆第一区调大队（乌鲁木齐）（图 8）。2004 年 4 月，工作站领导换届，第二届主任由田建荣（时任新疆地矿局局长）、孙革和莫斯布鲁格共同担任，左学义和宋松山（时任新疆第一区调大队队长）任常务副主任，德国阿什拉夫及吉林大学孙跃武教授任副主任。工作站管委会新委员还包括德方的怡德（Eder J.）、艾特尔（Eitel B.）、弗莱茨纳（Pfretzschner H.-U.）、泰恩（Thein J.）及马丁（Martin T.）教授等。2009 年 9 月，工作站第三届领导班子会议在乌鲁木齐举行，新增补新疆 305 办公室主任王宝林和德方波恩大学地质古生物所所长马丁教授为副主任（图 9）。

中德合作新疆地质工作站是中国新疆地质系统第一个与德国合作建立的国际性合作研究机构。在中德双方共同努力下，工作站在合作科研、人才培养及学术交流等方面发挥了重要作用，取得了一批重要科研成果，在提高新疆区域地质调查水平乃至扩大找寻新疆中生代石油及煤炭等化石能源远景等方面发挥了重要作用。工作站运行期间，新疆第一区调大队的专家和管理人员曾多次访问德国，并派员进修，德国科学家及师生多次来新疆工作及实习。工作站由此成了中德科学合作的"服务接待站"。工作站的工作一直持续到 2013 年，不仅将中德地质合作推向高潮，也加速了新疆区调地质队伍的国际化进程。与此同时，工作站也成为中德联合新疆科研队（以下简称"中德科研队"）的重要组成部分。

自 2000 年起，中德新疆地学合作得到中国国家基金委（NSFC）、德国科学基金会（DFG）、中德科学中心（SGSPC）、新疆自然资源厅及新疆地矿局等单位的大力支持。20 多年的时间里，中德合作新疆工作站及中德科研队在新疆共举行 3 次大型国际学术研讨会和多次野外专题联合科学考察，来自中德两国及美、日等国家和地区的 20 余所大学及研究机构的专家参与；出版会议论文集 3 部，发表论文百余篇，在国际上产生广泛影响。2005 年，在中

**图 8　中德合作新疆地质工作站**

Fig. 8　Sino-German Co-Working Station for Geosciences in Xinjiang (SGWSGX)

A. 中德合作新疆地质工作站办公楼（2000~2006，乌鲁木齐）；B. 成立仪式前，新疆维吾尔自治区领导（左3）会见工作站负责人（2000）；C. 中德科研队在准噶尔野外（2001）；D. 早侏罗世地层野外考察，右起：孙革，周汉忠，莫斯布鲁格，小李，李洁，王鑫甫，后排为麦什（1997）；E、F. 工作站接待来访的德、美专家座谈，其中，E 站立者为左学义大队长，F 站立者为图宾根大学校长赛西（2001）；G. 中德合作新疆项目入选中科院与德国马普学会优秀合作项目之一（2001）

A. Office of SGWSGX in Urumqi（2000–2006）；B. Before the ceremony for establishment of the SGWSGX, the leader of Xinjiang government（left 3）meeting the headers of the Station（2000）；C. Field work in the Junggar Basin（2001）；D. Field work on Lower Jurassic strata, from left: Wang X. F., Li J., Young Li, Mosbrugger V., Zhou H. Z., Sun G., the back row Maisch M.,（1997）；E, F. The Station welcoming the German and American guests: the standing in E, the Director of RGSX, Prof. Zuo X. Y.; the standing in F, Prof. Schaich E., President of Univ. Tuebingen, Germany（2001）；G. The Sino-German cooperative project selected as one of the excellent projects of the Program of the CAS–Max–Planck Soc.（2001）

图 9 活跃的中德新疆科学合作

Fig. 9 Sino-German active scientific cooperation in Xinjiang

A、B. 中德新疆国际学术会议，时任新疆维吾尔自治区副主席依明巴海（A 前排中）、中德科学中心常务副主任陈乐生（A 前排右 5、B 中）及李廷栋院士（A 前排左 6）等出席（2004，乌鲁木齐）；C、D. 室内外学术交流（2004）；E. 中德塔里木野外考察（2007）；F. 中德中心领导参观吉林大学地质博物馆（2005）；G. 中德新疆工作站第三届合作协议签字仪式，莫斯布鲁格（左 2）及孙革（左 1）等出席（2009，乌鲁木齐）

A, B. Sino-German symposium in Xinjiang, Vice-Governor, Prof. Arken Imirbaki （mid），Academician Li T. D. （left 6） and Prof. Chen L. S. （right 5） in presence （Urumqi, 2004）；C, D. Scientific exchanges in door and in field; E. Field work of Sino-German scientific team in Tarim （2007）；F. The leaders of SGSPC visiting Geological Museum of Jilin Univ. （2005）；G. Signing for the 3rd term of SGWSGX, Profs. Mosbrugger （left 2） and Sun G. （left 1） in presence （Urumqi, 2009）

德科学中心的大力支持下，中德科研队在吉林大学建立了首个中德古生物与地质联合实验室，中德双方共有 13 个科研单位参加，由孙革与莫斯布鲁格共同担任实验室主任，其工作重点仍是继续开展中国新疆的地质古生物合作研究等。2005 年 10 月，在吉林大学隆重举行中德古生物与地质联合实验室成立仪式，时任中德科学中心主任（中方）韩建国、吉林大学校长周其凤院士等应邀出席。

多年来，随着中德两国科学家对新疆中生代地质古生物的合作研究的展开，包括在新疆多次联合举办国际学术会议，中德两国的地学合作范围也扩大到美国、日本等国家的高校及科研机构。与此同时，中德两国年轻的研究生和博士也在新疆的合作研究中得到锻炼和提高，经包括联合培养博士生等合作，一批年轻的学者学术上得到迅速成长。2001 年，中德合作新疆地质工作站的合作入选中科院 – 德国马普学会合作优秀项目（图 8，G）；2010 年 10 月，应中德科学中心之邀，孙革和莫斯布鲁格代表中德科研队在北京出席"中德科学中心成立十周年庆典"，并在庆典大会上做了唯一的专题汇报。

# 第 2 章

# 新疆中生代地层掠影

　　新疆的中生代地层研究历史久远，多年来也取得众多的研究成果。本书报道的新疆中生代地层概述是中德科研队 20 多年来合作研究成果的最新总结，其中特别囊括了科研队在新疆准噶尔盆地和吐哈盆地中生代生物地层及其化石研究方面的新发现、新进展，如在准噶尔盆地郝家沟剖面首次建立起具高分辨率的 $T_3$–$J_2$ 孢粉地层序列，首次发现郝家沟晚三叠世植物群，首次在硫磺沟发现侏罗纪哺乳动物新组合（包括"准噶尔兽"等），首次在吐哈盆地发现迄今最大的侏罗纪恐龙"新疆巨龙"、侏罗纪恐龙足迹群及"新疆龟"新类群等，对准噶尔盆地中侏罗世白杨河植物群及吐哈盆地沙尔湖植物群首次进行了深入研究。上述新发现及研究新进展对推动新疆中生代盆地生物演化、地层对比、古生态环境演化乃至石油及煤等沉积矿产找寻等研究，均具有十分重要的意义。

## 2.1　三大盆地概况

　　新疆中生代地层分布广泛，主要集中在准噶尔、塔里木和吐哈三个盆地（图 1，D；图 10）。塔里木盆地面积最大（约 53 万平方千米），准噶尔盆地位居第二（约38万平方千米），吐哈盆地面积最小（约4.8万平方千米）。天山山脉将准噶尔盆地与塔里木盆地南北分隔，吐哈盆地则位于天山东段延续的山麓之间，呈东西向展布。吐哈盆地煤田的煤炭储量（沙尔湖煤田）居全国前列，塔里木盆地和准噶尔盆地分布的主要是油田。

新疆中生代三大盆地均形成于晚古生代褶皱构造带之上，其基底多为晚古生代地层，在许多地区（如准噶尔盆地南缘大龙口至玛纳斯－红沟一线）中生代早三叠世地层甚至与晚二叠世顶部的地层呈连续沉积，为研究二叠系—三叠系（P–T）界线创造了有利条件。新疆三大盆地的构造类型均属于"多类型叠加复合型盆地"。中生代时期，各盆地的沉降中心曾不断迁移，如准噶尔盆地，三叠纪时期盆地沉降中心曾位于乌鲁木齐以北，侏罗纪时期迁移到昌吉－玛纳斯河一带，而白垩纪沉降中心又迁移到盆地中部。就构造运动强度的差异而言，中生代时期，塔里木盆地是西强东弱，准噶尔盆地是南强北弱，而吐哈盆地是北强南弱（康玉柱，2003）。由于中生代时期三大盆地曾遍布河、湖和沼泽，动植物异常丰富，形成了丰富的石油、煤及天然气等矿藏，为中国化石能源的储备与开发提供了宝贵的财富。

由于上述三个盆地在中生代形成与演化的地质背景、时间以及生物与沉积特征均基本一致，地层也基本上可以对比；又由于准噶尔盆地中生代地层更为发育，化石更为丰富，研究程度相对更高；因此，本书以准噶尔盆地的地层为主要标准，初步形成一个统一的中生代地层简表（表1）。

**图 10　新疆三大中生代盆地中生代地层示意图**

Fig. 10　Sketch map showing the Mesozoic strata in the three basins of Xinjiang

A. 准噶尔盆地吉木萨尔大龙口早三叠世地层出露地；B. 阜康三工河早—中侏罗世地层；
C. 塔里木盆地库车早侏罗世含煤地层；D. 吐哈盆地哈密白垩纪地层

A. Lower Triassic in Dalongkou of Jimusar, Junggar Basin; B. Lower and Middle Jurassic strata in
Sangonghe section of Junggar Basin; C. Lower Jurassic coal-bearing strata in Kuqa of Tarim Basin;
D. Lower Cretaceous from Hami in Turpan-Hami Basin

**表 1 新疆中生代地层简表（以准噶尔盆地为例）**

Table 1 Simplified chart of Mesozoic strata of Xinjiang sampled with Junggar Basin

| | | | 地层单元 | | 主要岩性 | 备注 |
|---|---|---|---|---|---|---|
| 白垩系 K | 上 | K₂ | 下紫泥泉子组 | | 灰绿、红褐色砂岩，底部白色砂砾岩 | 产介形类等化石 |
| | | | 东沟组 | | 紫红色砂砾岩为主 | 乌尔禾等地产恐龙（塔里木西南部为海相） |
| | 下 | K₁ | 吐谷鲁群 | 连木沁组 | 杂色砂泥岩为主 | 产双壳类等化石 |
| | | | | 胜金口组 | 灰绿灰黄色砂泥岩为主 | 产鱼化石 |
| | | | | 呼图壁组 | 紫色砂泥岩为主 | 产双壳类等化石 |
| | | | | 清水河组 | 灰绿黄绿色砂泥岩，底部为砾岩 | 产准噶尔翼龙，鹦鹉嘴龙 |
| 侏罗纪 J | 上 | J₃ | 喀拉扎组 | | 紫红色砂砾岩为主 | （在吐哈盆地产蝴蝶龙） |
| | | | 齐古组 | | 紫色杂色砂岩为主 | 产新疆巨龙 |
| | 中 | J₂ | 头屯河组 | | 黄绿色细粉砂岩为主 | 产恐龙足迹群 |
| | | | 西山窑组 | | 灰色含煤砂岩粉砂岩为主 | 产植物化石，巨厚煤层 |
| | 下 | J₁ | 三工河组 | | 黄色砂岩为主，夹砾岩 | 产植物化石 |
| | | | 八道湾组 | | 灰色砂岩粉砂岩含煤层，底部为灰白色砂砾岩 | 含煤层，产植物化石 |
| 三叠系 T | 上 | T₃ | 郝家沟组 | | 灰色含煤砂岩粉砂岩为主 | 产植物及鲨化石 |
| | | | 黄山街组 | | 灰黄色砂砾岩为主 | 产植物化石 |
| | 中 | T₂ | 克拉玛依组 | | 灰绿色细砂粉砂岩为主 | 产鱼化石 |
| | 下 | T₁ | 上仓房沟群 | 烧房沟组 | 杂色砂砾岩为主 | 产孢粉及介形类化石 |
| | | | | 韭菜园组 | 紫红色砂砾岩为主 | 产水龙兽化石 |

## 2.2　新疆中生代地层

新疆中生代陆相地层特征较鲜明：（1）三叠纪。早三叠世以红色或杂色沉积为主，中三叠世以黄绿色细碎屑岩组成的湖泊相沉积为主，晚三叠世以灰色及黄灰色河流－湖泊相沉积为主，其晚期在局部地区成煤。（2）侏罗纪。早—中侏罗世以灰色含煤沉积为主，含大量工业煤层，中—晚侏罗世以杂色或紫色沉积为主，含大量恐龙化石；晚侏罗世晚期以紫色沉积为主。（3）白垩纪。早白垩世基本上为灰黄－灰绿色沉积与杂色沉积交互出现，早白垩世最早期沉积缺失，晚白垩世以黄灰色及紫色沉积为主，富含恐龙等化石（图10，图 11）。

此外，自白垩纪中期起至晚白垩世，新疆塔里木盆地西部尚有少量海相地层存在（席党鹏等，2019）。

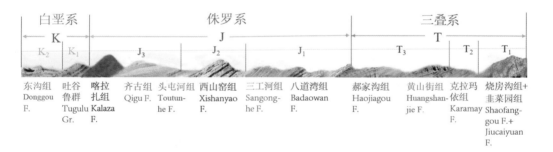

图 11　新疆中生代地层示意图（以准噶尔盆地为例）

Fig. 11　Sketch illustration of Mesozoic strata in Xinjiang shown by the sample in Junggar Basin

### 2.2.1　三叠系

#### （1）下三叠统

新疆早三叠世陆相地层主要为红色或杂色沉积，产水龙兽（*Lystrosaurus*）等脊椎动物化石。下三叠统与下伏上二叠统为整合接触，其界线地层典型剖面（即 P–T 界线剖面）在准噶尔盆地的吉木萨尔镇以南的大龙口出露较好。

　　大龙口剖面位于乌鲁木齐以东的阜康县吉木萨尔镇南约 15 千米的大龙口村的南部、河东岸有天然露头展示。该 P–T 界线代表了显生宙以来最重要的生物大灭绝事件之一（图 12，图 13）。

　　据新疆地矿局（1993）、李永安等（2003）及本科研队（Ashraf et al., 2004）等研究，大龙口 P–T 界线剖面显示的综合地层层序如下：

**上覆地层：**　　　　　　　　　　　　　　　　　　　　　　　　　　　　　　**厚度（米）**

5. 烧房沟组（$T_{1s}$）：紫色砂岩、粉砂岩，产孢粉等化石 ·························· 260~311

4. 韭菜园组（$T_{1j}$）：紫色粉砂质泥岩夹黄绿色粉砂岩，
   产孢粉及水龙兽（*Lystrosaurus*）等脊椎动物化石 ······················ 220~282

**界线地层：**

3. 锅底坑组（P–T）：灰绿色、紫色粉砂质泥岩夹灰黑色
   及杂色粉砂岩及泥岩，产植物（孢粉）、瓣鳃、介
   形类及叶肢介等化石·································· 144

**下伏地层：**

2. 梧桐沟组（$P_{3w}$）：灰色及灰黑色砂砾岩夹粉砂岩及
   泥岩，产孢粉等 ·································· 220

1. 泉子街组（$P_{3q}$）：紫色及灰绿色粉砂岩、泥岩夹砂
   砾岩，产植物 *Callipteris–Comia–Inopteris* 组合及瓣
   鳃类等化石·································· 122~242

　　锅底坑组（P–T）为 P/T 界线地层：之上为韭菜园子组（$T_{1j}$），之下为梧桐沟组（$P_{3w}$）；梧桐沟组之下的泉子街组（$P_{3q}$）产典型的晚二叠世安哥拉植物群的 *Callipteris–Comia–Inopteris* 植物组合化石，以及瓣鳃、脊椎动物等化石。有关 P–T 界线的具体划分目前仍在深入研究中（图 12，图 13）。

**（2）中三叠统**

　　新疆中三叠世地层主要以克拉玛依组为代表，主要为灰绿色、黄色砂岩，夹泥岩和砾岩等湖泊相沉积，底部通常有厚约 3.8 米的细砾岩，全组总厚 265~885 米，富含植物、双壳类、叶肢介、鱼类及其他脊椎动物等化石。建组剖面在准噶尔盆地西北部的克拉玛依附近，盆地东南部的阜康县小泉沟等地

图 12　吉木萨尔大龙口 P-T 界线及其附近地层（据 Ashraf et al.，2004）

Fig. 12　The P-T boundary strata in Dalongkou section of Jimsar (after Ashraf et al., 2004）

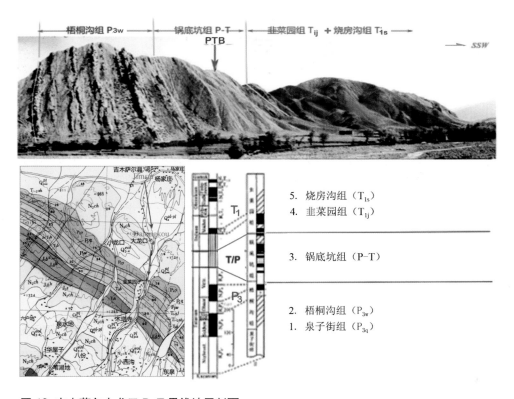

图 13　吉木萨尔大龙口 P-T 界线地层剖面

Fig. 13　The P-T boundary strata in Dalongkou of Jimusar with their simple descriptions

有该组典型剖面。该组与下伏烧房沟组（$T_{1s}$）及上覆黄山街组（$T_{3hs}$）均为整合接触（图14）。

　　该组产植物化石 *Neocalamites carcinoides*、*N. carrerei*,、*Bernoullia zeilleri*、

**图 14　新疆准噶尔盆地中三叠世克拉玛依组**

Fig. 14　Middle Triassic Karamay Formation in Junggar Basin

A、B. 吉木萨尔大龙口中三叠世克拉玛依组远望；C. 克拉玛依组剖面；D. 在克拉玛依组野外工作，左起：吴绍祖，阿什拉夫，孙革，李洁（1997）；E. 克拉玛依组底界界碑

A，B. Looking over the Karamay Formation（$T_2$）；C. Section showing the Karamay Formation; D. Field work in the Karamay Formation（from left）: Prof. Wu S. Z., Ashraf A. R., Sun G., and Li J.（1997）；E. Boundary marker showing the base of Karamay Formation

*Daenaeopsis fecunda*、*Chiropteris? yuani*、*Glossophyllum shensiense* 和 *Thinnfeldia nordenskioldi* 等，双壳类 *Ferganoconcha sibirica*，*F. rotunda* 和 *Sibiriconcha anodontites*，叶肢介 *Mesolimnadiopsis karamaica* 和 *Lioestheria euenkiensis*，鱼类 *Sinosemionodus urumuchi* 等，以及其他脊椎动物 *Parakannemeyeria brevirostris* 和 *Parotosaurus* sp. 等（新疆地矿局，1993）。

### （3）上三叠统

新疆晚三叠世地层主要为河流相或河流 – 湖泊相沉积，以准噶尔盆地为例，自下而上以黄山街组（T$_{3hs}$）和郝家沟组（T$_{3hj}$）为代表，典型剖面为乌鲁木齐市以西约 50 千米的郝家沟剖面。晚三叠世早 – 中期的黄山街组（T$_{3hs}$）主要由黄灰色河流相沉积组成，该组典型剖面在阜康县的小泉沟，为灰色及灰绿色泥岩夹砂质泥岩及粉砂岩，富含鱼化石 *Bogdania fragmenia*、*Fukangichthys longidorsa* 和 *Fukangolepis barbavos* 等。在郝家沟剖面，黄山街组主要为黄灰色砂岩、粉砂岩及少量砾岩，厚约 153 米，产植物化石拟丹尼蕨 – 南漳叶（*Danaeopsis-Nanzhangophyllum*）组合。黄山街组在这里反映河流相沉积可能与郝家沟地区更接近准噶尔盆地的南部或西南部边缘有关。

整合沉积于黄山街组之上，是晚三叠世晚期的郝家沟组（T$_{3hj}$ 剖面第 20~41 层）。郝家沟剖面是该组的建组剖面，命名于 1959 年（新疆地矿局，1993），1997 年，本书作者在这里实测了地层剖面，结果显示郝家沟组厚约 280 米（Sun et al.，2001b）。郝家沟组由湖泊相及部分河流相沉积组成，主要包括灰色砂岩及粉砂岩，以及少量砾岩，其上部夹多层炭质粉砂岩及煤层或煤线，所产植物化石主要以舌叶 – 准苏铁果（*Glossophyllum-Cycadocarpidium*）组合为代表（Sun et al.，2001b，2010）。该组顶部灰色 – 灰黑色粉砂岩与早侏罗世八道湾组（J$_{1b}$）底部的灰白色砂砾岩连续沉积，为研究 T–J 界线提供了有利条件（图 15）。

图 15 郝家沟剖面的晚三叠世地层

Fig. 15 The Upper Triassic strata in Haojiagou geological section

A. 郝家沟组（$T_{3hj}$）与八道湾组（$J_{1b}$）；B. 郝家沟组上覆八道湾组底部的灰白色砂砾岩；C、D. 郝家沟组含煤地层，其中 C 示科研队野外工作（右起：李洁，周汉忠，王鑫甫。1997）；E~G. 晚三叠世黄山街组：E 为黄山街组起点，F 为黄山街组中部，G 为含丰富植物化石层 A. The Haojiagou Formation（$T_{3hj}$）and Badaowan Fm（$J_{1b}$）; B. The basal grayish white conglomerate and sandstone of Baodaowan Fm; C、D. Coal−bearing strata of Haojiagou Fm（$T_{3hj}$），（C. from right: Li J., Zhou H. Z. and Wang X. F., 1997）; E−G. Huangshanjie Fm（$T_{3hs}$）: E showing the beginning of the formation, F showing the beds yielding abundant plant fossils, G showing the middle part of the formation

### 2.2.2 侏罗系

新疆侏罗纪地层主要为河湖相沉积：早、中侏罗世主要为以灰色调为主的砂岩、粉砂岩夹灰黑色巨厚的含煤沉积；晚侏罗世主要为以紫红色为主的砂岩、粉砂岩夹杂色砂岩沉积（图 16，图 17）。根据本书作者的实际工作，参考新疆地矿局、邓胜徽等及王思恩等的资料（新疆地矿局，1993；邓胜徽等，2010，2015；王思恩等，2012），以准噶尔盆地为例，侏罗系地层自下而上为：

**（1）下侏罗统**

八道湾组（$J_{1b}$）是以灰色砂岩为主的含煤沉积，底部为灰白色砂砾岩，主要反映河流相与沼泽相，厚约 622 米，含 *Dictyophyllum* 及锥叶蕨（*Coniopteris*）等植物化石，以及 *Unio* 及 *Ferganoconcha* 等双壳类化石。八道湾组的典型剖面在乌鲁木齐以东的八道湾煤矿，但在郝家沟剖面也有较完好出露，其中，特别是该组底部的灰白色砾岩及砂岩颇具特征（图 16，A~C）。

三工河组（$J_{1s}$）是以灰黄色、杂色砂岩为主的含湖泊相沉积，局部地区见有厚层的灰黄色砾岩层；全组厚约 882 米，含植物及瓣鳃类等化石；建组剖面在阜康县三工河，在建组剖面上本书作者曾采得 *Equisetites laterale*（Phill.）Morris 和膜蕨型锥叶蕨（*Coniopteris hymenophylloides* Brongniart）等植物化石，以往的报道还发现有双壳类 *Ferganoconcha*、*Sibiriconcha* 和 *Unio*，以及叶肢介 *Palaeolimnadia baitianbaensis* Chen，及 *Euestheria tenuiformis*（Zaspelova）等（Chen et al.，1998）（图 16，D、E）。

**（2）中侏罗统**

中侏罗世地层是新疆主要含煤地层，以著名的西山窑组（$J_{2s}$）为代表。西山窑组以灰色和灰黄色砂岩、粉砂岩为主，含巨厚含煤沉积，总厚约 980 米，主要显示湖泊 – 沼泽相。西山窑组建组剖面在乌鲁木齐西郊的西山窑煤矿；典型剖面在准噶尔盆地西北部的托里白杨河、乌鲁木齐以西的玛纳斯河等地；在吐哈盆地，典型剖面以沙尔湖煤田剖面为代表，该煤田是中国最大的煤田之一。西山窑组盛产植物等化石，著名的中侏罗世白杨河植物群及沙尔湖植物群均产于该组（图 17，A~F）。

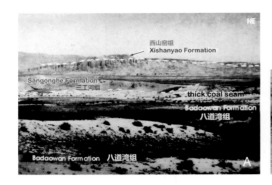

图 16 新疆早侏罗世地层

Fig. 16 Lower Jurassic strata in Xinjiang

A~C. 早侏罗世八道湾组（郝家沟剖面），其中 B、C 示八道湾组底部灰白色砂砾岩；D. 八道湾组（右侧）与上覆三工河组（黄色岩层）；E. 早侏罗世三工河组建组剖面（阜康三工河）；F. 早侏罗世含煤地层（塔里木盆地库车）

A~C. Lower Jurassic Badaowan Formation（from Haojiagou section）: B, C showing the grayish white conglomerate and sandstone; D. Badaowan Formation（right）and its overlying Sangonghe Formation（yellow beds）; E. Lower Jurassic Sangonghe Formation（from Sangonghe section of Fukang）; F. Lower Jurassic coal-bearing strata in Kuqa of Tarim Basin

　　头屯河组（$J_{2t}$）位于西山窑组之上，以灰绿、灰黄和灰紫色的湖泊相杂
色泥岩为主，厚约654米，含植物、介形类、瓣鳃类及叶肢介等化石；建组剖
面在乌鲁木齐以西的头屯河。准噶尔盆地东部奇台将军庙一带的石树沟组下
部可能相当于本组，富含硅化木等化石；吐哈盆地的三间房组可能也属于本组，
其中已发现大型恐龙足迹群等化石（图17，G~I）。据邓胜徽等研究，头屯
河组的同位素测年为166.2~160.8 Ma，主体沉积时期为中侏罗世晚期（邓胜徽
等，2015）。

**图 17　新疆中侏罗世地层**

Fig. 17　Middle Jurassic strata in Xinjiang

A~C. 白杨河中侏罗世西山窑组（准噶尔盆地）；D~F. 沙尔湖中侏罗世西山窑组，沙尔湖
煤田（吐哈盆地）；G. 奇台将军庙中侏罗世晚期石树沟组（准噶尔盆地）；H、I. 鄯善七
克台中侏罗世晚期三间房组（吐哈盆地）

A~C. Middle Jurassic Xishanyao Formation in Baiyanghe section（Junggar Basin）；D~F.
Xishanyao Formation in Shaerhu Coal-Mine（Tu-Ha Basin）；G. Middle Jurassic Shishugou
Formation（Junggar Basin）；H, I. Middle Jurassic Sanjianfang Formation in Shanshan（Tu-Ha
Basin）

（3）上侏罗统

齐古组（$J_{3q}$）以紫红色夹绿色砂岩、泥岩为主的湖泊相沉积，厚约683米，建组剖面在准噶尔盆地西北部呼图壁县的齐古。齐古组是新疆最盛产恐龙等脊椎动物的层位；在准噶尔盆地奇台将军庙曾发现体形庞大的中加马门溪龙（*Mamenchisaurus sinocanadorum*），本科研队在硫磺沟发现哺乳动物准

**图 18 新疆晚侏罗世地层**

Fig. 18 Upper Jurassic strata in Xinjiang

A~D. 晚侏罗世齐古组（准噶尔盆地奇台将军庙），其中B和C示科研队野外恐龙化石研究（2004）；E~G. 晚侏罗世齐古组（吐哈盆地鄯善），E和G示科研队恐龙发掘，F为该组发现的新疆巨龙化石（2012）

A–D. Upper Jurassic Qigu Formation in Jiangjunmiao area of Qitai（Junggar Basin）: B and C showing the investigation on dinosaurs by the Sino–German research team（2004）; E–G. Excavations of dinosaurs by the research team in Shanshan, F showing the dinosaur *Xinjiangotitan*（2012）

噶尔兽（*Dsungarodon*），在吐哈盆地鄯善发现鄯善新疆巨龙（*Xinjiangotitan shanshanensis*）及规模宏大的鄯善恐龙足迹群和乌苏新疆龟（*Xinjiangchytes wusu*）等脊椎动物化石（图 18）。有关齐古组的地质年代，王思恩与高林志曾发表他们在准噶尔盆地石河子以南（沙湾县境内）红沟剖面测得的齐古组锆石 SHIMP U–Pb 年龄为 164.6 ± 1.4 Ma（王思恩等，2012），此后，邓胜徽等在准噶尔盆地侏罗纪多重地层研究中报道，齐古组顶部年龄值为 155.3 Ma，沉积时间约 5.5 Ma, 包括牛津阶（Oxfordian）的上部和启莫里阶（Kimmeridgian）的中下部（邓胜徽等，2015）。

　　喀拉扎组（$J_{3k}$）是上覆于齐古组的一套以灰褐色砾岩夹砂岩为主的河流相沉积，厚 50~800 米，在地面上多展现为高山地貌。该组建组剖面在乌鲁木齐附近昌吉市的喀拉扎山，有孢粉等化石报道。在准噶尔盆地，喀拉扎组的顶部被早白垩世吐谷鲁群的清水河组不整合覆盖，邓胜徽等认为该不整合时间超过 7 Ma（新疆地矿局，1993；邓胜徽等，2015；张驰等，2021）（图 19）。

**图 19 准噶尔盆地晚侏罗世卡拉扎组**

Fig. 19 Upper Jurassic Kalaza Formation from Junggar Basin

### 2.2.3 白垩系

白垩纪地层在新疆三大盆地均有分布,但早白垩世最早期地层大都缺失。据新疆地矿局研究,准噶尔盆地的早白垩世地层称吐谷鲁群(土古里克群),主要为灰绿灰黄色沉积及灰、绿、紫等杂色沉积,总厚约 927.5 米。该群被划分为 4 个组,自下而上分别为连木沁组(杂色砂泥岩为主)、胜金口组(灰绿灰黄色砂泥岩为主)、呼图壁合组(紫色砂泥岩为主)、清水河组(灰绿黄绿色砂泥岩,底部为砾岩)。吐谷鲁群以湖泊相沉积为特征,含介形类、鱼、恐龙及翼龙等化石(新疆地矿局,1993)。翼龙化石以著名的准噶尔翼龙(*Dsungaripterus*)为代表,其模式种"魏氏准噶尔翼龙"(*D. weii*)由中国著名古脊椎动物学家杨钟健(1964)研究与命名,是中国发现的第一具较完整的翼龙化石。该翼龙头部具尖长锥状上翘、无齿的吻端,以及圆钝短粗的牙齿、厚重的骨骼,翼间窝具双拱形后缘,吻端发育类似鸟类的角质喙,推测其具有挖掘捕食泥沙中具有坚硬外壳的贝壳类等食物的生活习性。数十年后,汪筱林等在吐哈盆地的哈密地区又发现新的翼龙化石——哈密翼龙(*Hamipterus*)(Wang et al.,2014,2017)(图 20,A~C)。

吐谷鲁群的时代曾长期认为属早白垩世晚期[阿普特期—阿尔布期(Aptian–Albian)](新疆地矿局,1993)。近年来,据杨景林等及席党鹏等研究,认为该群时代近属于整个早白垩世:清水河组为贝里阿斯晚期至瓦兰今早期(late Berriasian–early Valanginian),呼图壁组为瓦兰今晚期至欧特里夫期(late Valanginian–Hauterivian),胜金口组为巴雷姆期至阿普特早期(Barremian–early Aptian),连木沁组为阿普特中期—阿尔布中期(middle Aptian–middle Albian)(席党鹏等,2019;杨景林等,2008)。

新疆晚白垩世地层在准噶尔盆地主要由东沟组和下紫泥泉子组组成。东沟组以灰紫色、杂色砂砾岩为主,厚约 884.5 米。在准噶尔盆地北部乌尔禾(魔鬼城)至三个泉一带,曾发现恐龙等化石(图 20,H)。郑秀亮等根据介形类化石认为,该组时代可能为康尼亚克期—坎潘期(Coniacian–Campanian)(郑秀亮等,2013;席党鹏等,2019)。东沟组之上的紫泥泉子组的时代及

**图 20　新疆白垩纪地层**

Fig. 20　Cretaceous strata in Xinjiang

A~C. 哈密早白垩世吐谷鲁群及其翼龙骨骼化石；D~G. 准噶尔翼龙，其中 D 和 E 示头骨化石，F 和 G 示复原图；H. 晚白垩世东沟组（乌尔禾）；I. 晚白垩世嘉裕龙（*Chiayusaurus*）；J. 晚白垩世鄯善龙（*Shanshanosaurus*）（I 和 J 自吐哈盆地）（A~G 据 Wang et al., 2014, 2017）

A–C. Lower Cretaceous Tugulu Group and its pterosaur bone fossils; D–G. *Dzungaripterus*: D and E showing the skull, F and G as its constructions; H. Upper Cretaceous Donggou Formation in Wurhe （Junggar Basin）; I. *Chiayusaurus*（$K_2$）; J. *Shanshanosaurus*（$K_2$）; I and J both from Tu–Ha Basin（A–G after Wang et al., 2014, 2017）

地层划分目前仍存在争议：介形类和轮藻化石研究认为，该组下部（即下紫泥泉子组）为晚白垩世，上部为古近纪，中间可能存在地层缺失（杨景林等，2012；郑秀亮等，2013）；杨景林等及席党鹏等认为，至少下紫泥泉子组的时代属于坎潘晚期至马斯特里赫特期（late Campanian–Maastrichtian）（杨景林等，2012；席党鹏等，2019）。

此外，据董枝明及赵喜进等研究，新疆晚白垩世地层富含恐龙化石，包括在吐哈盆地晚白垩世地层曾发现鄯善龙（*Shanshanosaurus*）及嘉裕龙（*Chiayusaurus*）等（图 20）（董枝明，1973，1977；赵喜进，1980）。

值得提及的是，据席党鹏等（2020）研究，在白垩纪中、晚期，塔里木盆地西部为一个喇叭状、向西开口的海湾，该海湾属于东特提斯洋的一个分支。塔里木盆地白垩纪海相或海陆过渡相地层（自下而上）分别为早白垩世克孜勒苏群（巴雷姆期—阿尔布期），晚白垩世库克拜组、乌依塔克组、依格孜牙组［塞诺曼期—马斯特里赫特期（Cenomanian–Maastrichtian）］，以及吐依洛克组（K–Pg 过渡）。上述地层富含有孔虫、介形类、沟鞭藻、孢粉、双壳类、腹足类，以及少量菊石、腕足类、海胆和鲨鱼牙齿等化石。塔里木盆地的海侵始于早白垩世晚期（阿普特晚期—阿尔布早期，相当于克孜勒苏群中上部沉积期），但规模有限。大规模的海侵始于晚白垩世初期，从晚白垩世至古新世，共经历了 5 次大规模的海侵—海退，直至始新世（距今约 41 Ma）海水退出塔里木盆地南部的昆仑山山前；在大约 34 Ma，海水退出盆地北部的天山山前。上述海侵—海退事件可能受塔里木盆地区域构造运动和全球海平面变化的双重影响（席党鹏等，2020）（图 21）。

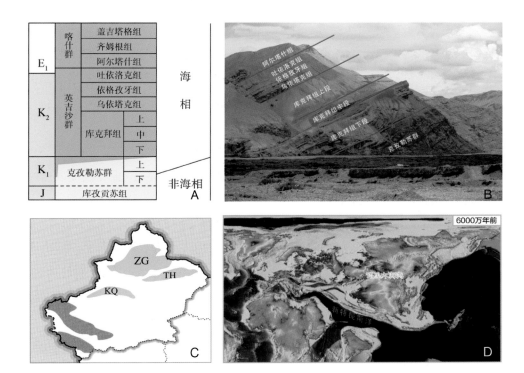

**图 21　西南塔里木盆地白垩纪地层及古地理**

**Fig. 21　Cretaceous strata and paleogeography of southwestern Tarim basin**

A. 塔里木盆地白垩纪地层简表；B. 白垩纪地层示意图；C. 新疆白垩纪海相与非海相地层分布示意图；D. 塔里木海湾示意图（A、B 和 D 据席党鹏等，2020；C 据席党鹏等，2019；本书均略有修改）

A. Cretaceous stratigraphic chart of Tarim Basin; B. Cretaceous strata; C. Sketch map showing the distributions of marine（blue colors） nd non-marine strata; D. Sketch illustration showing the Tarim Bay（A, B, and D after Xi et al., 2020; C. after Xi et al., 2019; with some revisions）

# 新疆中生代生物化石新发现

中德合作新疆科研队自 1997 年起对新疆中生代的合作研究开展了 20 余年，从准噶尔盆地开始，之后转向吐哈盆地，并涉猎部分塔里木盆地的幅域，取得了一系列重要地层古生物新发现。古生物新发现包括：早三叠世四足动物、中生代恐龙、翼龙、龟、鳄、鲨鱼、哺乳动物，以及植物（含孢粉）等（Martin et al., 2010b）。其中的主要代表性成果包括：首次发现郝家沟晚三叠世植物群、大龙口早三叠世水龙兽、晚侏罗世新的哺乳动物组合、侏罗纪新疆龟三个新种、迄今最大的侏罗纪恐龙——新疆巨龙及侏罗纪恐龙足迹群，以及首次开展新一轮中侏罗世白杨河植物群及沙尔湖植物群研究等。本书提及的中生代化石新发现均为本科研队取得的最新成果。

## 3.1 大龙口早三叠世水龙兽

### 3.1.1 大龙口的 P-T 界线之争

二叠纪与三叠纪之交（距今约 2.52 亿年时期），地球上发生了自显生宙以来一次重大的生物大灭绝事件，地球上 85% 的生物灭绝。自早三叠世（即中生代之初）起，新的生命又开始出现。研究这一时期的生物化石与地层对研究此间地球生物与环境的变化具有十分重要的意义。

准噶尔盆地的吉木萨尔大龙口有壮观的二叠系与三叠系（P/T）界线，其附近地层天然露头均完好出露（图 12）。P-T 界线地层为锅底坑组，其下伏地层梧桐沟组（$P_{3w}$）产晚二叠世孢粉化石，之下的泉子街组产典型的晚二叠世安加拉植物群（以 *Callipteris-Comia-Inopteris* 植物组合为代表）及瓣鳃类、脊椎动物等化石；上覆的早三叠世韭菜园组（$T_{1j}$）产著名的早三叠世四足类动物——水龙兽（*Lystrosaurus*）。由此，这里的 P-T 界线及其附近地层多年来已成为研究 P/T 界线地层的"热点"之一。20 世纪 80 年代，由中国地质科学院牵头的中美合作科研队曾在这里开展了有关 P/T 界线地层的合作研究，后因某些争论而中断。本次中德合作科研队的到来，使大龙口 P/T 界线研究焕发了新的生机，也倾注了中德两国科学家辛勤的汗水。

## 3.1.2 水龙兽的发现

水龙兽（*Lystrosaurus*）属于爬行动物下孔亚纲（Synapsida）二齿兽类（dicynodontid）。它头大，颈短，身体呈桶状，模样有点像今日的河马，大小与家猪相似。它鼻子向下弯曲，头骨和眼眶更高，身体结构有一些现生哺乳动物的进步特征，但头部仍然是原始的。除了两枚硕大的上犬齿外，下颌没有牙齿，有一个角质喙。根据对其肩部、腰带和四肢的分析，水龙兽有一种伸直的步态。它特殊的头骨形状被认为适应在干旱环境中掘洞或摄取坚硬的植物，强壮的前脚适合挖掘。骨组织学表明，其鼻孔内有瓣膜状结构，可能适于潜入水中捕捉食物，推测其一些种可能是半水生的。总体分析，水龙兽可能生活在潮湿的热带或温暖的森林地带的湖边，以植食为生（图 22，A、B）。

水龙兽主要生活在二叠纪晚期至三叠纪早期，在南非、印度、俄罗斯及中国新疆等地均有发现，通常被视为早三叠世的标准化石，也是新疆中生代早期生物化石的重要代表。本书提及的水龙兽是该属幼年个体的一具几乎完整的骨架（图 22，A），该化石是本科研队的两位青年专家麦什（Maisch M.）和马茨克（Matzke A.）2002 年在大龙口剖面早三叠世韭菜园组下部发现的（图

**图 22 新疆大龙口早三叠世水龙兽的发现**

Fig. 22 Discovery of Early Triassic *Lystrosaurus* from Dalongkou of Xinjiang

A. 早三叠世水龙兽化石；B. 水龙兽复原图；C. 化石发现者麦什博士（左）和马茨克博士在新疆野外；D. 化石发现地点（大龙口韭菜园组，$T_1$）；E. 在德国斯图加特自然史博物馆，右起：叶捷教授（IVPP），马茨克，麦什，孙革，崔宝琛博士（2000）；F. 德国队员在大龙口野外过河（2005）

A. Fossil *Lystrosaurus* from $T_1$; B. Reconstruction; C. The main discoverers Dr. Maisch（left）and Dr. Matzke in Xinjiang; D. Fossil site of *Lystrosaurus* from $T_1$ in Dalongkou section; E. Research team members in Stuttgart Museum（NH）, from right: Prof. Ye J.（IVPP）, Matzke A., Maisch M., Sun G., Dr. Cui B. C.（2000）; F. Crossing the river in Dalongkou by German scientists（2005）

22，C、D）。该标本是迄今中国保存最完整的水龙兽骨架化石之一，也是迄今在非洲以外发现的保存最完整的水龙兽幼年个体骨架。该化石以其个体较小、颅骨顶背部具轻微装饰，以及肩部和腰带未分化等为特征，其神经弓未与椎骨中央融合、腕关节和跗骨骨化很弱等也是其特征之一。除上述幼年个体骨架外，还发现其他几个头骨及颅后骨骼碎片等（Maisch et al.，2004）。

# 3.2　郝家沟晚三叠世植物群

郝家沟晚三叠世植物群的典型化石产地为乌鲁木齐以西的郝家沟剖面（图15），该植物群是作者于 1997~1999 年在郝家沟剖面开展的系统研究的成果，包括植物大化石和孢粉化石两部分。植物大化石研究成果于 2001 年起开始发表（Sun et al.，2001b，2010），孢粉化石研究成果自 1998 年开始正式发表（Ashraf et al.，1998，2001，2004，2010）。

## 3.2.1　郝家沟晚三叠世大化石植物群

郝家沟植物群的大化石包括了两个植物组合，分别为：（1）黄山街植物组合［拟丹尼蕨 - 南漳叶（*Danaeopsis-Nanzhangophyllum*）组合］，时代为晚三叠世卡尼期—诺利期（Carnian-Norian）（图 23）；（2）郝家沟植物组合［舌叶 - 准苏铁果（*Glossophyllum-Cycadocarpidium*）组合］，时代为诺利期—瑞替期（Norian-Rhaetian）（图 24）。郝家沟晚三叠世植物群的总体组成包括：

**有节类 Equisetales**

1. *Neocalamites* sp. 新芦木（未定种）

**真蕨类 Filicinae**

2. *Todites?* sp. 似托第蕨？（未定种）

3. *Danaeopsis tenuinervis* Sun, Mosbrugger et Li 细脉拟丹尼蕨

4. *Danaeopsis* sp. 拟丹尼蕨（未定种）

5. *Cladophlebis grabauiana* Sze 葛利普枝脉蕨

图 23  郝家沟晚三叠世植物群黄山街组合

Fig. 23  The Huangshanjie Assemblage of Late Triassic Haojiagou flora

A. 新芦木（未定种）；  B. 葛利普枝脉蕨；C. 高氏枝脉蕨；D、E. 郝家沟南漳叶；F、G.
奇裂（？）丁菲羊齿；H. 短尖头带羊齿；I. 带羊齿（未定种）；J. 细脉拟丹尼蕨；比例
尺 =1 cm（据 Sun et al.，2001b，2010）

A. *Neocalamites* sp.; B. *Cladophlebis grabauiana*; C. *C. kaoiana*; D, E. *Nanzhangophyllum
haojiagouense*; F, G. *Thinfeldia? minisecta*; H. *Taniopteris mucronulata*; I. *Taniopteris* sp. J.
*Danaeopsis tenuinervis*; bar=1 cm（after Sun et al., 2001b, 2010）

6. *Cladophlebis kaoiana* Sze 高氏枝脉蕨

7. *Cladophlebis* sp. 枝脉蕨（未定种）

**种子蕨类 Pterispermae**

8. *Thinfeldia? minisecta* Sun, Mosbrugger et Li 奇裂？丁菲羊齿

9. *Thinnfeldia* sp. 丁菲羊齿（未定种）

10. *Nanzhangophyllum haojiagouense* Sun，Mosbrugger et Li 郝家沟南漳叶

**苏铁类？Cycadophytes？**

11. *Taniopteris mucronulata* Sun，Mosbrugger et Li 短尖头带羊齿

12. *Taniopteris* sp. 带羊齿（未定种）

**银杏类 Ginkgoales**

13. *Ginkgoidium* sp. 准银杏（未定种）

14. *Sphenobaiera?* sp. 楔拜拉（未定种）

15. *Glossophyllum shensiense* Sze 陕西舌叶

16. *Solenites haojiagouensis* Yang et al. 郝家沟似管状叶

**松柏类 Coniferales**

17. *Cycadocarpidium erdmanni* Nathorst 爱德曼准苏铁果

18. *Cycadocarpidium swabii* Nathorst 斯瓦比准苏铁果

19. *Cycadocarpidium* sp. 准苏铁果（未定种）

**（1）晚三叠世黄山街植物组合**

黄山街组植物组合（拟丹尼蕨－南漳叶组合）产于黄山街组下部的灰黄色－黄灰色粉砂岩之中。主要组成分子包括新芦木（未定种）（*Neocalamites* sp.）、细脉拟丹尼蕨（*Danaeopsis tenuinervis* Sun, Mosbrugger et Li）、拟丹尼蕨（未定种）（*Danaeopsis* sp.）、葛利普枝脉蕨（*Cladophlebis grabauiana* Sze）、高氏枝脉蕨（*C. kaoiana* Sze）、奇裂（？）丁菲羊齿（*Thinfeldia? minisecta* Sun，Mosbrugger et Li）、郝家沟南漳叶（*Nanzhangophyllum haojiagouense* Sun，Mosbrugger et Li）、短尖头带羊齿（*Taniopteris mucronulata* Sun，Mosbrugger et Li）及带羊齿（未定种）（*Taniopteris* sp.）等（图23）。

上述组合中，拟丹尼蕨（*Danaeopsis*）是中国晚三叠世植物群常见分子，也是中国北方晚三叠世植物群的主要代表性分子；葛利普枝脉蕨及高氏枝脉蕨是中国北方晚三叠世延长植物群重要组成分子；南漳叶（*Nanzhangophyllum*）是中国南方晚三叠世植物群常见分子，以往主要见于湖北南漳等地的晚三叠世地层。因此，黄山街组植物组合明显显示了晚三叠世植物群的面貌（Sun et al., 2001b, 2010）。

### （2）晚三叠世郝家沟植物组合

郝家沟植物组合（舌叶–准苏铁果组合）主要产于郝家沟组的浅灰色–灰色粉砂岩夹煤线的岩层中。主要组成分子包括似托第蕨？（未定种）（*Todites?* sp.）、枝脉蕨（未定种）（*Cladophlebis* sp.）、丁菲羊齿（未定种）（*Thinnfeldia* sp.）、准银杏（未定种）（*Ginkgoidium* sp.）、楔拜拉？（未定种）（*Sphenobaiera?* sp.）、陕西舌叶（*Glossophyllum shensiense* Sze）、郝家沟似管状叶（*Solenites haojiagouensis* Yang et al.）、爱德曼准苏铁果（*Cycadocarpidium erdmanni* Nathorst）、斯瓦比准苏铁果（*C. swabii* Nathorst）和准苏铁果（未定种）（*Cycadocarpidium* sp.）等。

上述植物组合中，陕西舌叶是中国北方晚三叠世植物群的主要代表性分子；爱德曼准苏铁果及斯瓦比准苏铁果是中国南方晚三叠世植物群常见分子，并主要见于晚三叠世瑞替期（Rhaetian）。因此，郝家沟组植物组合也显示了晚三叠世植物群的面貌，且较之黄山街组植物组合时代略年轻（Sun et al., 2001b, 2010）（图 24）。

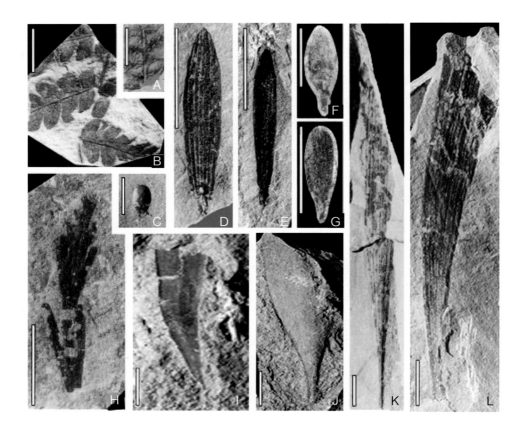

**图 24 郝家沟晚三叠世植物群郝家沟组合**

Fig. 24 The Haojiagou Assemblage of Late Triassic Haojiagou flora

A. 似托第蕨？（未定种）； B. 枝脉蕨（未定种）； C. 石籽（未定种）； D、E. 斯瓦比准苏铁果；F、G. 爱德曼准苏铁果；H. 丁菲羊齿（未定种）；I. 楔拜拉？（未定种）；J. 准银杏（未定种）；K、L. 陕西舌叶；比例尺 =1 cm（据 Sun et al.，2001b，2010）

A. *Todites*? sp.; B. *Cladophlebis* sp.; C. *Carpolithus* sp.; D, E. *Cycadocarpidium swabii*; F, G. *Cycadocarpidium erdmanni*; H. *Thinnfeldia* sp.; I. *Sphenobaiera*? sp.; J. *Ginkgoidium* sp.; K, L. *Glossophyllum shensiense*；bar=1 cm（after Sun et al., 2001b, 2010）

## 3.2.2 郝家沟晚三叠世孢粉植物群

除植物大化石外，本植物群还发现大量孢粉化石。据本科研队研究，晚三叠世孢粉植物群至少由 57 属 93 种组成。其中，蕨类植物孢子占 53.97%，裸子植物花粉占 46.03%。晚三叠世黄山街组孢粉组合主要以 *Concavisporites–Duplexisporites problematicus–Lophotrites sangburensis–Cyclotrites oligografinifer* 组合为代表，时代主要显示诺利期（Norian）；晚三叠世郝家沟组孢粉组合主要以 *Concavisporites–Duplexisporites problematicus–Ruccisporites tuberculatus* 组合为代表，时代主要显示瑞替期（Rhaetian）。上述两组合共同显示了晚三叠世孢粉植物群的面貌（Ashraf et al.，1998，2001，2004，2010）。主要孢粉组成分子如下（表 2，图 25，图 26）。

表 2 郝家沟晚三叠世植物群孢粉组成简表 （据 Ashraf et al.，1998，2001）

Table 2 Composition of the Late Triassic Haojiagou palynoflora (after Ashraf et al., 1998, 2001)

| 序号 | 孢粉分类群<br>Taxa of sporopollen | 黄山街组<br>$T_{3hs}$ | 郝家沟组<br>$T_{3hj}$ | 中文学名<br>Scientific name in Chinese |
|---|---|---|---|---|
| | 苔藓类 Bryophytes / 蕨类 Pteridophytes | | | |
| 1 | *Stereisporites perforatus* | | X | |
| 2 | *Lycopodiacidites* sp. | X | | 石松孢（未定种） |
| 3 | *Aratrisporites fischeri* （Klaus）Playford et Dettmann, 1965 | | X | 费歇尔离层单缝孢 |
| 4 | *Aratrisporites scabratus* Klaus, 1960 | | X | 粗糙离层单缝孢 |
| 5 | *Aratrisporites strigosus* Playford，1965 | | X | 薄壁离层单缝孢 |
| 6 | *Aratrisporites* sp. | | X | 离层单缝孢（未定种） |
| 7 | *Densoisporites velatus* Weyland et Krieger, 1953 | X | | 膜缘拟套环孢 |
| 8 | *Osmundacidites wellmanii* Couper, 1953 | X | | 威氏紫萁孢 |
| 9 | *Osmundacidites wellmanii conatus* | X | | 威氏紫萁孢（变种） |
| 10 | *Osmundacidites* sp. | | X | 紫萁孢（未定种） |
| 11 | *Todisporites concentricus* Li，1981 | X | | 同心托第蕨孢 |
| 12 | *Auritulinasporites intrastriatus* | X | | 间纹厚唇孢 |

（续表）

| 序号 | 孢粉分类群<br>Taxa of sporopollen | 黄山街组<br>T$_{3hs}$ | 郝家沟组<br>T$_{3hj}$ | 中文学名<br>Scientific name in Chinese |
|---|---|---|---|---|
| 13 | *Concavisporites bohemiensis* Thiergart，1953 | X | | 波希米亚凹边孢 |
| 14 | *Concavisporites mortoni*（Jersey）Jersey，1962 | | X | 莫氏凹边孢 |
| 15 | *Concavisporites toralis*（Leschik，1955）Nilsson，1958 | | X | 膨胀凹边孢 |
| 16 | *Concavisporites* sp. | X | | 凹边孢（未定种） |
| 17 | *Dictyophyllidites mortoni*（Jersey）Playford et Dettmann, 1965 | X | X | 莫氏网叶蕨孢 |
| 18 | *Dictyophyllidites* sp. | | | 网叶蕨孢（未定种） |
| 19 | *Annulispora folliculusa*（Rogalska）De Jersey，1959 | X | | 大圈环圈孢 |
| 20 | *Apiculatisporis* sp. | X | X | 锥瘤孢（未定种） |
| 21 | *Baculatisporites comanmensis*（Cookson）Potonie，1956 | X | | 科茅姆棒瘤孢 |
| 22 | *Carnisporites granulatus* | X | | 粒纹卡尼孢 |
| 23 | *Camarozonosporites rudis*（Leschik）Klaus, 1960 | X | | 皱纹楔环孢 |
| 24 | *Cyclogranisporites* sp. | X | | 圆形粒面孢（未定种） |
| 25 | *Duplexisporites gyrates* Playford et Dettmann，1965 | X | | 圆形旋脊孢 |
| 26 | *Duplexisporites problematicus* | X | | 可疑旋脊孢 |
| 27 | *Duplexisporites scanicus* | | X | 旋脊孢 |
| 28 | *Duplexisporites* sp. | X | | 旋脊孢（未定种） |
| 29 | *Habrozonosporites decorates* Lu et Wang，1980 | X | | 装饰丽环孢 |
| 30 | *Habrozonosporites* sp. | X | | 丽环孢（未定种） |
| 31 | *Hsuisporites liaoningensis* Pu et Wu，1985 | X | | 辽宁徐氏孢 |
| 32 | *Kraeuselisporites reissingeri* | X | | 稀饰环孢 |
| 33 | *Limatulasporites haojiagouensis* | | X | 郝家沟背光孢 |
| 34 | *Limatulasporites* sp. | | X | 背光孢（未定种） |
| 35 | *Polycingulatisporites simplex* | | X | 简单多环孢 |
| 36 | *Polycingulatisporites* sp. | X | | 多环孢（未定种） |
| 37 | *Punctatisporites* sp. | X | | 圆形光面孢（未定种） |
| 38 | *Reticulatisporites* sp. | X | | 粗网孢（未定种） |

（续表）

| 序号 | 孢粉分类群 Taxa of sporopollen | 黄山街组 T<sub>3hs</sub> | 郝家沟组 T<sub>3hj</sub> | 中文学名 Scientific name in Chinese |
|---|---|---|---|---|
| 39 | *Retusotriletes mesozoicus* Klaus, 1960 | X | | 中生弓脊孢 |
| 40 | *Retusotriletes* sp. | | X | 弓脊孢（未定种） |
| 41 | *Verrucosisporites* sp. | X | | 圆形块瘤孢（未定种） |
| | **裸子植物 Gymnospermae** | | | |
| 42 | *Abietineaepollenites* sp. | | X | 冷杉粉（未定种） |
| 43 | *Alisporites australis* De Jersey, 1962 | X | | 澳大利亚阿里粉 |
| 44 | *Alisporites thomasii*（Couper）Pocock, 1962 | X | | 托马斯阿里粉 |
| 45 | *Alisporites* sp. | | X | 阿里粉（未定种） |
| 46 | *Chasmatosporites apertus*（Rogalska）Nilsson, 1958 | | X | 无盖广口粉 |
| 47 | *Chasmatosporites anagrammensis* | | X | 阿纳格拉姆广口粉 |
| 48 | *Chasmatosporites hianu* Nilsson, 1958 | X | | 敞开广口粉 |
| 49 | *Chasmatosporites verruculosus* Qian, Zhao et Wu, 1983 | X | | 瘤纹广口粉 |
| 50 | *Chasmatosporites* sp. | | | 广口粉（未定种） |
| 51 | *Ginkgocycadophytus* nitidus (Balme) Pocock | X | | 光泽银杏苏铁粉 |
| 52 | *Cycadopites fragilis* Singh, 1964 | X | | 脆弱苏铁粉 |
| 53 | *Cycadopites dilucidus*（Bolkh.）Zhang W. P., 1984 | X | | 清晰苏铁粉 |
| 54 | *Cycadopites reticulatus*（Nilsson）Arjang, 1975 | X | | 网纹苏铁粉 |
| 55 | *Cycadopites rugugranulatus* Jiang et Du, 2000 | X | | 皱粒苏铁粉 |
| 56 | *Cycadopites subgranulosus*（Couper）Bharadwaj et Singh, 1964 | X | | 亚颗粒苏铁粉 |
| 57 | *Cycadopites tivoliensis* De Jersey, 1971 | X | X | 蒂沃利苏铁粉 |
| 58 | *Cycadopites* sp. | | X | 苏铁粉（未定种） |
| 59 | *Ovalipollis* sp. | X | | 卵形粉（未定种） |
| 60 | *Piceites expositus* | X | | 显露拟云杉粉 |
| 61 | *Piceites* sp. | X | | 拟云杉粉（未定种） |
| 62 | *Pinuspollenites* sp. | X | | 松粉（未定种） |
| 63 | *Striatoabietites aytugii* | X | | 冷杉多肋粉 |
| 64 | *Platysaccus lopsinensis* | X | | 洛普辛蝶囊粉 |

（续表）

| 序号 | 孢粉分类群<br>Taxa of sporopollen | 黄山<br>街组<br>T₃ₕₛ | 郝家<br>沟组<br>T₃ₕⱼ | 中文学名<br>Scientific name in<br>Chinese |
|---|---|---|---|---|
| 65 | *Platysaccus* sp. | X | | 蝶囊粉（未定种） |
| 66 | *Paleoconiferus asaccatus* Bolk., 1956 | X | | 无囊古松柏粉 |
| 67 | *Podocarpites multisimus* （Bolkh.）Pocock, 1962 | | X | 多凹罗汉松粉 |
| 68 | *Podocarpites* sp. | | X | 罗汉松粉（未定种） |
| 69 | *Protopinus* sp. | X | | 原始松粉（未定种） |
| 70 | *Psophosphaera* | X | | 皱球粉（未定种） |
| 71 | *Cordaitina* sp. | X | | 科达粉（未定种） |
| 72 | *Florinites* sp. | X | | 弗洛林粉（未定种） |
| 73 | *Pseudocrustaesporites* sp. | X | | 假克鲁氏粉（未定种） |
| 74 | *Tetrasaccus* sp. | X | | 四囊粉（未定种） |
| 75 | *Quadraeculina limbata* Maljavkina, 1949 | | X | 真边四字粉 |
| 76 | *Chordasporites singulichorda* | X | | 辛古里克达单脊粉 |
| 77 | *Chordasporites* sp. | | X | 单脊粉（未定种） |
| 78 | *Lunatisporites rhaeticus* | X | X | 瑞替四肋粉 |
| 79 | *Protohaploxypinus* sp. | X | | 单束多肋粉（未定种） |
| 80 | *Taeniaesporites combinatus* s Qu et Wang, 1990 | X | X | 联结宽肋粉 |
| 81 | *Taeniaesporites xingxianensis* Qu, 1984 | X | | 兴县宽肋粉 |
| 82 | *Taeniaesporites leptocorpus* Qu, 1984 | | X | 薄体宽肋粉 |
| 83 | *Taeniaesporites* sp. | X | X | 宽肋粉（未定种） |
| 84 | *Vittatina* sp. | X | | 叉肋粉（未定种） |
| | **大孢子 Megaspores** | | | |
| 85 | *Echitriletes prerussus* | | X | 前俄刺面大孢 |
| 86 | *Otyniosporites* sp. | | X | 奥氏大孢（未定种） |
| 87 | *Aneuletes* sp. | | X | 无缝大孢（未定种） |
| 88 | *Cabochonicus carbunculus* | | X | 疖痈卡波大孢 |
| 89 | *Cabochonicus* sp. | | X | 卡波大孢（未定种） |
| 90 | *Horstisporites* sp. | | X | 地垒孢（未定种） |
| 91 | *Bothriotriletes* sp. | | X | 双耳三缝孢（未定种） |
| 92 | *Paxillitriletes* sp. | | X | 帕氏三缝孢（未定种） |
| 93 | *Calamospora rhaeticus* | | X | 瑞替芦木大孢 |

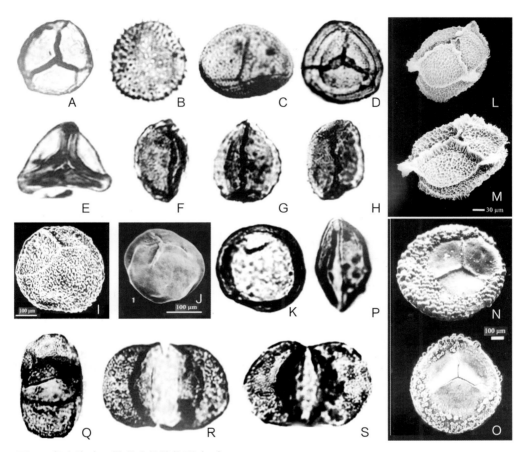

**图 25　郝家沟晚三叠世孢粉植物群（1）**

Fig. 25　The sporopollen from Upper Triassic Huangshanjie and Haojiagou formations（1）

黄山街组孢粉：A. 区域弓脊孢；B. 科茅姆棒瘤孢；C. 威氏紫箕孢；P. 光泽银杏苏铁粉；R. 澳大利亚阿里粉；Q. 小阿里粉。郝家沟组孢粉：D. 郝家沟背光孢；E. 膨胀凹边孢；F–H. 薄壁离层单缝孢；I. 奥氏大孢（未定种）；J. 瑞替芦木大孢；K. 无盖广口粉；L、M. 哈里斯穴网大孢；N、O. 疖痈卡波大孢；S. 多凹罗汉松粉（据 Ashraf et al.，2001）

From Huangshanjie Formation: A. *Retusotriletes arealis*; B. *Baculatisporites comanmensis*; C. *Osmundacidites wellmanii*; P. *Ginkgocycadophytus nitidus*; R. *Alisporites australis*; Q. *Alisporites parvus*. From Huajiagou Formation: D. *Limatulasporites haojiagouensis*; E. *Concavisporites toralis*; F–H. *Aratrisporites strigosus*; I. *Otyniosporites* sp.; J. *Calamospora rhaeticus* K. *Chasmatosporites apertus*; L, M. *Horstisporites hariisii*; N, O. *Cabochonicus carbunculus*; S. *Podocarpites multisimus* （after Ashraf et al., 2001）

晚三叠世黄山街组孢粉组合以 *Concavisporites–Duplexisporites problematicus–Lophotrites sangburensis– Cyclotrites oligografinifer* 组合为代表，大体显示诺利期（Norian）。晚三叠世郝家沟组孢粉组合以 *Concavisporites–Duplexisporites problematicus–Lophotrites sangburensis–Ruccisporites tuberculatus* 为代表，大体显示瑞替期（Rhaetian）(Ashraf et al., 1998，2001，2004，2010)（图 25，图 26）。

就古生态特征而言，郝家沟晚三叠世孢粉植物群显示了如下特征：（1）该植物群的卷柏科（Selaginellaceae）、石松科（Lycopodiaceae）及紫萁科（Osmundaceae）等类群均是喜湿型植物；（2）中等喜湿型植物包括蚌壳蕨科（Dicksoniaceae）、双扇蕨科（Dipteridaceae）、马通蕨科（Matoniaceae）及海金沙科（Schizaeaceae）等，相对前述喜湿型植物而言，生长环境可能略偏干燥些；（3）中生型植物（mesophytic forms）包括苏铁类、本内苏铁类、银杏类及松柏类等，它们的出现表明这里的古环境曾相对略干燥些。喜湿型植物和中等喜湿型植物的混生表明，郝家沟晚三叠世植被还是存在一定的季节性变化或植物生长高度的不同，如松柏类通常为森林的高层植被，而喜湿型的蕨类生长在森林的低层，容易得到水分的供给。对于这一点，沉积学研究也提供了旁证。

相比较而言，黄山街组孢粉组合植物多样性相对略高(41 属 58 种)。其中，裸子植物花粉（20 属 31 种）占整个植物组合约 53%，显示了当时的湿度较低；松柏类和非松柏类花粉占优势，代表性的类群包括膜缘拟套环孢（*Densoisporites velatus*）、间纹厚唇孢（*Auritulinasporites intrastriatus*）、大圈环圈孢（*Annulispora folliculusa*）、粒纹卡尼孢（*Carnisporites granulatus*）、可疑旋脊孢（*Duplexisporites problematicus*）、稀饰环孢（*Kraeuselisporites reissingeri*）、澳大利亚阿里粉（*Alisporites australis*）、网纹苏铁粉（*Cycadopites reticulatus*）、卵形粉（*Ovalipollis*），以及原始松粉（*Protopinus*）等（Ashraf et al., 1998，2001）。郝家沟组孢粉组合植物多样性相对略低（37 属 52 种），其中，一些孢子达到了晚三叠世

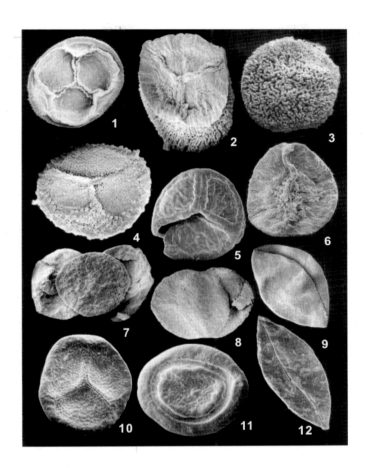

**图 26　郝家沟晚三叠世孢粉植物群（2）**

Fig. 26　The sporopollen from Upper Triassic Huangshanjie and Haojiagou formations（2）

1~3. 棒卡达孢；4. 粒纹卡尼孢；5. 皱纹楔环孢；6. 菲舍尔离层单缝孢；7. 双囊阿里粉；8. 编织假松粉；9. 苏铁粉（未定种）；10. 简单弓脊孢；11. 希拉特盘孢；12. 苏铁粉（未定种）。1~3、6、11 和 12 自郝家沟组，4、5、7、8 和 10 自黄山街组（据 Ashraf et al.，2004）

1–3. *Cardargasporites baculatus*; 4. *Carnisporites granulatus*; 5. *Camarozonosporites rudis*; 6. *Aratrisporites fisheri*; 7. *Alisporites bisaccus*; 8. *Psedopinus textilis*; 9. *Cycadopites* sp.; 10. *Retusotriletes simplex*; 11. *Discisporites psilatus*; 12. *Cycadopites* sp. Photos 1–3, 6, 11, and 12 from Haojiagou Formation; photos 4, 5, 7, 8, and 10 from Huangshanjie Formation（after Ashraf et al., 2004）

的高峰，如瑞替四肋粉（*Lunatisporites rhaeticus*）、瑞替芦木大孢（*Calamospora rhaeticus,*）、无盖广口粉（*Chasmatosporites apertus*）、可疑旋脊孢（*Duplexisporites problematicus*）及凹边孢（*Concavisporites*）等。从黄山街组至郝家沟组，孢粉类型有所增加，频繁的成煤过程可能表明，此间由于环境的沼泽化，供水已有些不足。相比之下，古气候可能更温和、潮湿，这一点植物大化石中也有显示。

郝家沟组顶部和早侏罗世八道湾组底部的交界可能是 T/J 界线。早侏罗世八道湾组的孢粉植物群与郝家沟晚三叠世孢粉植物群的面貌已明显不同。八道湾组的孢子离层单缝孢（*Aratrisporites*）和背光孢（*Limatulasporites*）已变得很少甚至完全缺失；裸子植物 "线囊型" 花粉（striate saccate pollen）也是如此（Ashraf et al.，1998，2001，2010）。

综合上述植物大化石及孢粉化石，郝家沟晚三叠世植物群总体上显示了与中国陕北延长植物群更为接近的面貌，如该植物群出现了较多的陕北延长植物群的组成分子拟丹尼蕨（*Danaeopsis*）、陕西舌叶（*Glossophyllum shensiense*）、葛利普枝脉蕨（*Cladophlebis grabauiana*）、高氏枝脉蕨（*C. kaoiana*），以及新芦木（*Neocalamites*）等，总体上属于中国北方晚三叠世拟丹尼蕨 – 舌叶（*Danaeopsis-Glossophyllum*）植物群。但与此同时，该植物群在植物组成上又出现了少量中国南方晚三叠世植物群的分子，如南漳叶（*Nanzhangophyllum*）、爱德曼准苏铁果（*Cycadocarpidium erdmanni*）及斯瓦比准苏铁果（*Cycadocarpidium swabii*）以及大量的带羊齿（*Taniopteris*）等。因此，郝家沟植物群显示了以中国北方晚三叠世植物群为主、兼具南方型植物群混生的植物群性质，它的南北方晚三叠世植物群混生特征可能与郝家沟植物群晚三叠世地处准噶尔盆地西南缘、距离中国晚三叠世南北方植物分区界线（孙革，1987，1993；孙革等，1995）不远有关。总之，郝家沟晚三叠世植物群的发现对于深入研究新疆晚三叠世植物群的分布与演化、中国晚三叠世植物分区等，具有较重要的意义。

# 3.3 侏罗纪新疆龟三个新种

侏罗纪龟化石广泛分布于中生代的新疆，以著名的新疆龟类（Xinjiangchelid）为代表。新疆龟科（Xinjiangchelidae）是俄罗斯古脊椎动物学家涅索夫（Nesov L.）（1990）所建，其代表属新疆龟（*Xinjiangchelys*）是中国古脊椎动物学家叶捷（1986）所建。2001~2012 年，本科研队在新疆准噶尔盆地硫磺沟和吐哈盆地鄯善先后发现大量龟化石，并在硫磺沟建立了两个新种，分别为周氏新疆龟（*Xinjiangchelys chowi*，发现于中侏罗世头屯河组）（Matzke et al.，2005）和齐古新疆龟（*X. qiguensis*，发现于晚侏罗世齐古组）（Matzke et al.，2004）；在吐哈盆地鄯善七克台，还发现新种——乌苏新疆龟（*X. wusu*，产于晚侏罗世齐古组）（Rabi et al.，2013；Wings et al.，2012）。

## 3.3.1 周氏新疆龟

周氏新疆龟（*Xinjiangchelys chowi* Matzke et al.，2005）是中德科研队马茨克等（Matzke et al.，2005）于乌鲁木齐以西约 40 千米的硫磺沟中侏罗世晚期头屯河组上部首次发现（图 27，A）。本种的主要特征为：形体大小中等，壳低而圆，甲壳至少长 34.5 厘米；第 1 椎骨最大，横向扩张；第 7 和第 8 椎骨间的背关节缺失；第 1 肋骨有游离侧肋端；第 2 肋骨侧缘呈三角形；至少在第 1/ 第 2 肋骨及第 2/ 第 3 肋骨之间有小的周边通道；具两个上臀骨，第 1 个明显大于第 2 个；臀骨短但明显加宽，腹侧有一内侧加厚的脊，向外延伸到第 2 个臀骨上；第 5 椎骨显著覆盖周围组织和臀骨；前外周，包括第 6 缘骨最前面的部分有沟槽；臀骨上的第 12 缘骨相当小；至少 2 个骶椎骨，只有 1 根骶肋骨；与甲壳松散连接的橡皮筋；甲壳、舌骨和下颌之间有较大的侧囟形通道；下颌骨后突短，外侧长凸起物突出，内侧部分厚度减小，沿正中缝有凸起（图 27，C）（Matzke et al.，2005）。

此种已发现完整的甲壳和右下胚层（hypoplastron）。它显示了新疆龟科和新疆龟属的重要特征为：第 1 椎骨比第 2 椎骨大，前外周被切除，后外周

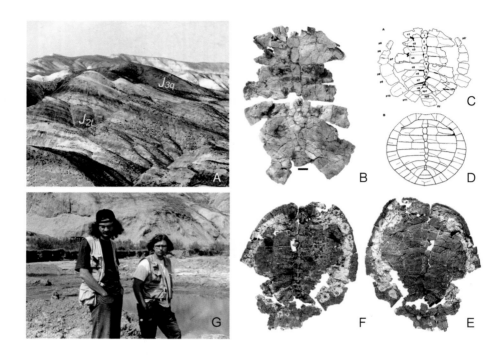

图 27　周氏新疆龟及齐古新疆龟

Fig. 27　*Xinjiangchelys chowi* and *X. qiguensis*

A. 硫磺沟侏罗纪头屯河组（$J_{2t}$）及齐古组（$J_{3q}$）；B~D. 周氏新疆龟（*Xinjiangchelys chowi*），中侏罗世头屯河组上部，其中 B 为化石龟甲图片，C 和 D 为甲板复原图（据 Matzke et al.，2005）；E、F. 齐古新疆龟（*X. qiguensis*）化石龟甲图片，晚侏罗世齐古组（据 Maisch et al.，2004）；G. 新疆龟两个新种的主要发现者：马茨克（右）和麦什

A. Turtle-bearing Middle Jurassic Toutunhe Formation and Upper Qigu Formation in Liuhuanggou; B-D. *Xinjiangchelys chowi* from Toutunhe Formation: B showing carapace, C and D showing reconstructions of the carapace（after Matzke et al., 2005）; E, F. *Xinjiangchelys qiguensis*: E showing carapace in dorsal view, F showing carapace in ventral view（after Maisch et al., 2004）; G. The main discoverers of the two new turtle species of *Xinjiangchelys*: Matzke A.( right ), and Maisch M.

扩张，第 1 胸肋缩小，以及质体桥的侧桩。本种区别于本属其他已知种的主
要特征为：第 1 肋骨具有游离肋骨端、至少两个甲壳前外侧外周凹门、一个
具有大侧凹门的薄橡皮筋和一条具有坚固销钉的正中下成形缝线。这些特征
表明，周氏新疆龟是迄今为止已知的最具衍生性的新疆龟；与天山新疆龟（*X.*
*tianshanensis*）一样，是迄今为止已知的最古老的新疆龟；表明新疆龟科可能
具有一个漫长的、复杂的演化体系（Matzke et al.，2005）。

### 3.3.2  齐古新疆龟

齐古新疆龟（*Xinjiangchelys qiguensis* Matzke et al.，2004）最早发现于乌
鲁木齐以西约 40 千米的硫磺沟剖面，晚侏罗世齐古组下部。该种也是新疆龟
类中具有最原始特征的种，主要特征为：其甲壳长度与周氏新疆龟相近；神
经节都有点不对称，第 7 和第 8 神经节间没有间隙。其甲壳的前外侧部分，
未发现游离肋骨末端和凹形通道；其甲壳最后面的边缘是向前弯曲的，而不
是像周氏新疆龟那样指向后面。齐古新疆龟暴露在第 8 肋腹内侧的骨盆关节
面较周氏新疆龟的明显要大。与周氏新疆龟相比，齐古新疆龟的胸甲与甲壳
融合得更紧密，且未见到外侧凹门或下胚层之间的"中楔"（medial pegs）。
与周氏新疆龟一样，下胚层内侧部分的厚度并没有减少（图 27，E、F）（Maisch
et al.，2004）。

### 3.3.3  乌苏新疆龟

本科研队在新疆中生代龟类化石发现中，以在吐哈盆地鄯善地区齐古组
发现的乌苏新疆龟（*Xinjiangchelys wusu*）保存最为完整，并以一龟化石集群
形式发现。这一新发现大大推动了对新疆龟骨骼形态学、古地理分布以及与
现生龟类演化关系等方面的认识。

乌苏新疆龟化石发现于鄯善七克台南戈壁滩、恐龙足迹化石点东北约 7
千米、新疆巨龙化石点西北约 5 千米的小山顶部（中心地理位置：42°57'31.5" N,
90°33'21.0" E）（图 28，A）。乌苏新疆龟（*Xinjiangchelys wusu*）化石产出地

层为晚侏罗世齐古组的砖红色及杂色砂岩所夹的灰色 – 灰绿色钙质砂岩及粉砂岩中。目前已发掘出乌苏新疆龟 3 个骨架化石，它们保存较完整，呈叠覆状。该种的主要特征：头骨较宽，腭骨和额骨更广泛参与眼眶形成，前额骨完全被额骨分隔，翼骨间隙发育，椎盾窄长。更为重要的是，该种发育一些原始颅底特征，如翼骨间隙和基蝶骨突，不同于现代龟类高骨化程度的颅底，表明在龟类演化中的生态适应方面的转变。这些特征也为探讨龟类的颈动脉系统的演化提供了依据。这些化石的发现，为详细了解新疆龟类化石骨骼特征提供了难得的完整骨架（图 28，图 29）。

该化石集群地的龟化石集中分布在 10 米 × 30 米的区域，核心区的龟化石保存密集，相互叠覆。完整的龟壳长约 20 厘米，龟化石的数量可达每平方米 36 只，按此计算，在 20 平方米范围内很可能发现有 720 枚龟化石保存。由此推测，该龟化石点的个体数量可达约 1800 只。此新发现将新疆侏罗纪时期的龟类化石数量扩充了一倍以上。

需要说明的是，此龟化石点外围区龟化石分布较零散，完整度也较差，这表明外围个体的死亡较早，埋藏前受到破坏，而核心区的个体可能死亡较晚，受破坏程度小。这一现象与一些现生水生龟类相似：在旱季时，随着水体的减少，龟类个体不断聚集，最后集中死亡。大量龟化石密集保存，可能由当时季节性的干旱造成龟群聚于水源地而形成。化石展示的龟类动物群的生态环境，可能与现今南美委内瑞拉（图 29，G）或澳大利亚穆瑞河等地的现生龟所处环境相似。该地区在侏罗纪中晚期的季节性干旱环境也进一步支持了对新疆龟生态环境的上述推断。

龟是一类奇特的爬行动物，身披甲壳，头、尾和四肢可缩入壳内。现生龟（Testudines）约有 330 种，广布于世界各大洲（除南极洲外），涉及陆地、河流和海洋等生境。依据脖颈收缩方式的不同，它们被分为隐颈龟（Cryptodira）和侧颈龟（Pleurodira）。隐颈龟种类最为丰富，约占总数的 75%，包括我们熟悉的鳖、陆龟、水龟、鳄龟、海龟等类群。侧颈龟相对较少，分为蛇颈龟和侧颈龟，分布仅限于南半球。现代龟类的起源和早期分异可追溯至侏罗纪时期。中国侏罗纪的龟类化石类型丰富、数量大、分异度高，是现代龟类起源和早期分异研究的热点之一。这些化石主要分布在西部的四川和新疆地区，以四川龟科（Sichuancheylidae）和新疆龟科（Xinjiangchelyidae）为主导。新疆地区的化石发现主要集中于准噶尔盆地和吐鲁番盆地，以四川龟科中的四川龟（*Sichuanchelys*）、新疆龟科的新疆龟（*Xinjiangchelys*）和安娜龟（*Annemys*）等为代表。它们个体为小型到中型，龟甲低平。其中，四川龟较为原始，发育有翼骨齿和中腹甲等，分布仅限于准噶尔盆地。新疆龟较为进步，缺失翼骨齿和中腹甲，与现生隐颈龟有着很近的亲缘关系，在两大盆地都有发现，

图 28　乌苏新疆龟及其产出地层

Fig. 28 *Xinjiangchelys wusu* and its bearing strata

A. 化石产地；B. 乌苏新疆龟；C-F. 野外工作及化石采集，其中 E 示化石野外打包；G. 化石修复

A. Turtle fossil site; B. *Xinjiangchelys wusu*; C-F. Field work for collecting fossils; G. Preparing fossils

A

A new xinjiangchelyid turtle from the Middle Jurassic of Xinjiang, China and the evolution of the basipterygoid process in Mesozoic turtles

Rabi et al.

An enormous Jurassic turtle bone bed from the Turpan Basin of Xinjiang, China

Oliver Wings, Márton Rabi, Jörg W. Schneider, Leonie Schwermann, Ge Sun, Chang-Fu Zhou & Walter G. Joyce

Naturwissenschaften
The Science of Nature

ISSN 0028-1042
Volume 99
Number 11

Naturwissenschaften (2012) 99:925-935
DOI 10.1007/s00114-012-0974-5

F

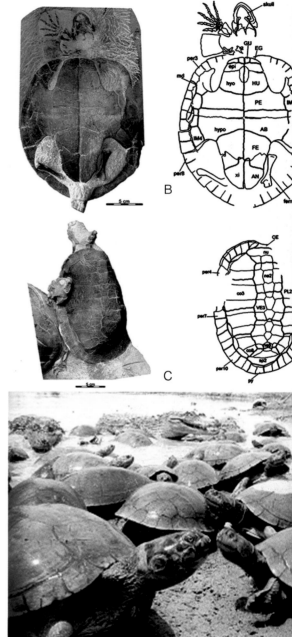

B

C

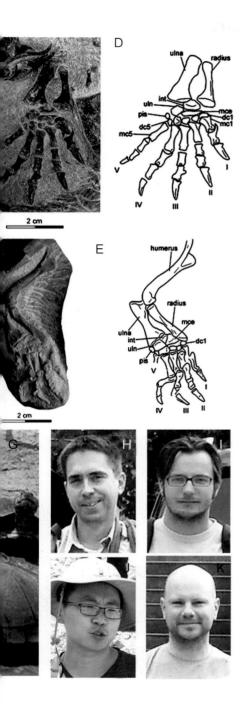

2 cm

2 cm

图 29　乌苏新疆龟

Fig. 29　*Xinjiangchelys wusu*

A. 三枚骨架照片（Rabi et al., 2013）；B、C. 头骨及甲壳（caparaces）；D. 前肢；E. 后肢（据 Rabi et al., 2013, 节选）；F. 德国《自然科学》杂志封面（Wings et al., 2012）；G. 与现生龟对比（示南美委内瑞拉龟群）；H~K. 发现乌苏新疆龟的青年专家（依次为：维恩斯，拉比，周长付，乔伊斯）

A. Skeletons of *Xinjiangchelys wusu*; B, C. Skulls and carapaces; D. Forelimb; E. Hind limbs（after Rabi et al., 2013; selected）; F. The published paper（2012）; G. Comparison with the living turtles and environments in Venezuela of South America; H–K. Four of the fossil discoverers（in order）: Wings, Rabi, Zhou, and Joyce

具有生物地层对比意义。过去，这些化石大都以龟甲保存，缺少完整骨架信息。本次新疆龟在鄯善地区晚侏罗世地层的新发现，无疑为研究新疆地区侏罗纪龟类化石的组成、分布与演化，以及其所处古生态环境等，提供了重要新资料，也为新疆地区侏罗纪地层的对比提供了龟类化石方面的支持。

乌苏新疆龟化石是由德国青年专家维恩斯（Wings O.）、乔伊斯（Joyce W.）、拉比（Rabi M.）和中国青年专家周长付等共同发现的，成果发表于国际学术刊物 *BMC*（2013）及德国的《自然科学》（*Naturwissenschaften*）（2012），曾引起国际学术界的轰动（图 29）。

# 3.4 白杨河中侏罗世植物群

## 3.4.1 白杨河植物群

一个世纪前，当俄罗斯地质学家奥勃鲁契夫初到白杨河采集化石（1904~1905，1909）、英国古植物学家秀厄德（Seward，1911）初步研究这里的植物化石时，准噶尔地区（包括白杨河）还是一片荒漠，受条件所限，化石采集和研究都很初步。如今，白杨河中侏罗世植物群的研究条件已大大改观，工作也不断深入。

白杨河植物群产于托里白杨河河谷（中心地理位置：84° 56′ E，46° 22′ N）两侧的中侏罗世西山窑组含煤地层（图 30，A、B）。秀厄德（1911）报道在这里发现的植物化石仅 16 属 21 种，包括 *Equisetites ferganensis* Sternberg，*Raphaelia diamensis* Seward、*Coniopteris hymenophylloides* Brongniart、*C. quinqueloba*（Phillips）Seward，*Eboracia lobifolia*（Phill.）Thomas，*Sphenopteris modesta* Leckenby，*Cladophlebis* sp.，*Ginkgo obrutschewi* Seward，*Baiera lindleyana* Schimper，*Phoenicopsis angustifolia* f. media Krasser，*Podozamites lanceolatus*（L. et H.）Braun，*P.* cf. *pulchellus* Heer，*Pityophyllum* cf. *staratschini* Heer 和 ?*Sphenolepidium* sp. 等（Seward，1911）。

为了搞清白杨河植物群的最初研究情况，本书作者之一的孙革曾多次来到俄罗斯圣彼得堡的俄罗斯地质博物馆（VSEGEI），这里收藏了俄罗斯地质学家奥勃鲁契夫当年采集，后来经过英国古植物学家秀厄德鉴定的化石标本和地质图件（图 30，C，D）。迄今，作者在白杨河发现的植物化石已超过 28 属 46 种；除丰富了对秀厄德当年报道的奥勃鲁契夫银杏（*Ginkgo obrutschewi*）、膜蕨型锥叶蕨（*Coniopteris hymenophylloides*）、费尔干似木贼（*Equisetites ferganensis*）、狄阿姆拉法尔蕨（*Raphaelia diamensis*）和裂叶爱博拉契蕨（*Eboracia lobifolia*）等几个重要的分类群的认识外，还对秀厄德有关 *Coniopteris quinqueloba* 和 *Sphenopteris modesta* 等分类群进行了修订，并详细研究了奥勃鲁契夫银杏的表皮构造、膜蕨型锥叶蕨的原位孢子及其囊群；新建了准噶尔楔拜拉（*Sphenobaiera jungarensis* Sun et Miao）、白杨河斯卡布果（*Scarburgia baiyangheensis* Yang et al.）等新的分类群；首次确认了费尔干杉（*Ferganella*）、贝加尔茨康叶〔*Czekanowskia*（*Vachrameevia*）*baikalica*〕、疏松薄果穗（*Leptostrobus laxifolia*）、槲寄生（*Ixostrobus*），以及似叶状体（*Thallites*）及新芦木（*Neocalamites*）等类群在白杨河植物群的存在（Sun et al.，2001b，2010，2021；孙革等，2006；苗雨雁，2005，2006，2017；杨涛等，2017，2018），对提高白杨河中侏罗世植物群的研究程度发挥了重要作用。

目前，作者已在白杨河发现的植物化石主要有（图 31~ 图 33）：

**苔藓类（Bryophytes）**

1. 似叶状体（未定种）*Thallites* sp.

**有节类（Equisetales）**

2. 费尔干似木贼 *Equisetites ferganensis* Seward

3. 似木贼（未定种）*Equisetites* sp.

4. 霍尔新芦木 *Neocalamites hoerensis*（Schimp.）Halle

**真蕨类（Filicinae）**

5. 布列亚锥叶蕨 *Coniopteris burejensis*（Zal.）Seward

6. 膜蕨型锥叶蕨 *C. hymenophylloides*（Brongn.）Harris

7. 裂叶爱博拉契蕨 *Eboracia lobifolia*（Phill.）Thomas

**图 30 白杨河植物群研究**

Fig. 30 The study of Baiyanghe flora

A、B. 白杨河及其地质剖面；C. 俄罗斯地质博物馆；D. 孙革在俄罗斯地质博物馆查阅地质图（2007）；E~H. 苗雨雁率队在白杨河剖面工作（2005）；I~L. 孙革率队在白杨河剖面工作（2015）

A, B. The Baiyang（Diam）River and its geological section; C. Russian Geological Museum（VSEGEI）in St. Petersburg; D. Sun G. viewing the geological map in VSEGEI （2007）; E-H. Field work in Baiyanghe headed by Miao Y. Y. （2005）; I-L. Field work in Baiyanghe headed by Sun G. （2015）

8. 山西枝脉蕨 *Cladophlebis shansiensis* Sze

9. 柴达木枝脉蕨 *C. tsaidamensis* Sze

10. 枝脉蕨（未定种）*Cladophlebis* sp.

11. 狄阿姆拉法尔蕨 *Raphaelia diamensis* Seward

12. 拉法尔蕨（未定种）*Raphaelia* sp.

**本内苏铁类（Bennettitales）**

13. 尼尔桑蕉羽叶（未定种）*Nilssoniopteris* sp.

**茨康类（Czekanowsiales）**

14. 狭叶拟刺葵 *Phoenicopsis angustifolia* Heer

15. 华丽拟刺葵 *P. speciosa* Heer

16. 坚直茨康叶 *Czekanowskia rigida* Heer

17. 贝加尔茨康叶 *C.*（*Vachrameevia*）*baikalica* Kirichkova et Samylina

18. 疏松狭轴穗 *Leptostrobus laxifolia* Heer

19. 狭轴穗（未定种）*Leptostrobus* sp.

20. 槲寄生（未定种）*Ixostrobus* sp.

**银杏类（Ginkgoales）**

21. 林德拜拉 *Baiera lindleyana* Schimper

22. 奥勃鲁契夫银杏 *Ginkgo obrutschewi* Seward

23. 胡顿银杏？ *G. huttonii* Harris？

24. 银杏（未定种）*Ginkgo* sp.

25. 准噶尔楔拜拉 *Sphenobaiera jungarensis* Sun et Miao

26. 清晰狭轴穗 *Stenorachis lepida*（Heer）Seward

**松柏类（Coniferas）**

27. 松型叶（未定种）*Pityophyllum* sp.

28. 裂鳞果（未定种）*Schizolepis* sp.

29. 柏型枝？（未定种）*Cupressinocladus ?* sp.

30. 落羽杉？（未定种）*Taxodium ?* sp.

**图 31 白杨河中侏罗世植物群的苔藓类及蕨类化石**

Fig. 31 Fossil mosses and ferns from Middle Jurassic Baiyanghe flora

A. 似叶状体；B~D. 费尔干似木贼，其中 C 为叶鞘，B 左边为横膈膜，D 为茎；E. 似木贼（未定种），示叶鞘；F. 霍尔新芦木；G~L. 膜蕨型锥叶蕨，其中 G 为营养叶，H 为生殖叶，I、J 为孢子囊堆，K 为孢子，L 为孢子囊群；M. 裂叶爱博拉契蕨；N、O. 狄阿姆拉法尔蕨；P. 拉法尔蕨（未定种）

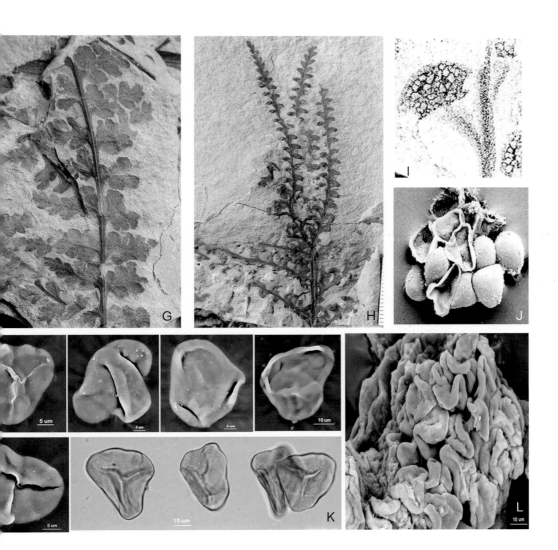

A. Thallites sp.; B–D. *Equisetites ferganensis*: C showing leaf sheath, B left showing diaphragm, D showing stem; E. *Equisetites* sp., showing leaf sheath; F. *Neocalamites hoerensis*; G–L. *Coniopteris hymenophylloides*: G showing sterile leaf, H showing fertile leaf, I and J showing sori, K showing spores, L showing spolongia; M. *Eboracia lobifolia*; N, O. *Raphaelia diamensis*; P. *Raphaelia* sp.

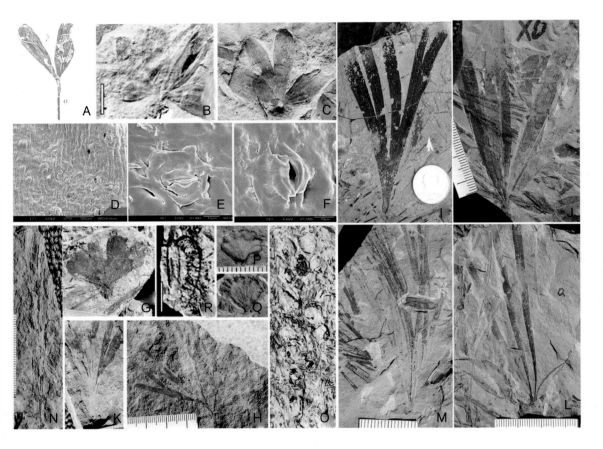

**图 32 白杨河中侏罗世植物群的茨康类和银杏类**

Fig. 32 Czekanowskiales and Ginkgoales in Middle Jurassic Baiyanghe flora

A~F. 奥勃鲁契夫银杏，其中 A 为本种的模式标本，B 及 C 为作者新采集的标本，D~F 为作者新研究的本种的表皮构造；G. 胡顿银杏？；H. 林德拜拉；I. 准噶尔楔拜拉；J、K. 华丽拟刺葵；L. 狭叶拟刺葵；M. 贝加尔茨康叶；N. 坚直茨康叶；O. 疏松狭轴穗；P、Q. 狭轴穗（未定种）；R. 槲寄生（未定种）

A~F. *Ginkgo obrutschewi*; G. *huttonii?*; H. *Baiera lindleyana*; I. *Sphenobaiera jungarensis*; J, K. *Phoenicopsis speciosa*; L. *Phoenicopsis angustifolia*; M. *Czekanowskia* (*Vachrameevia*) *baikalica*; N. *Czekanowskia rigida*; O. *Leptostrobus laxifolia*; P, Q. *Leptostrobus* sp.; R. *Ixostrobus* sp.

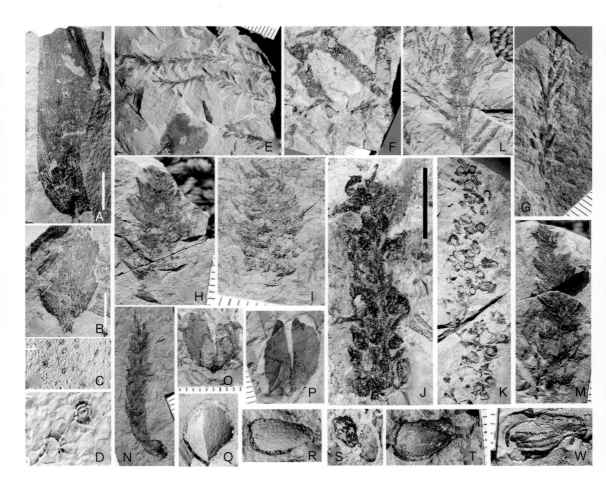

**图 33　白杨河中侏罗世植物群的松柏类**

Fig. 33　Conifers of Middle Jurassic Baiyanghe flora

A~D. 费尔干杉（未定种），其中 A、B 为叶，C、D 示下表皮气孔不定向排列；E. 短叶枞型枝；
F. 柏型枝？（未定种）；G. 落羽杉？（未定种）；H、I. 卵形似枞？；J、K. 白杨河斯卡布
果；L. 枞型枝（未定种 1）；M、N. 似果穗（未定种 1、2）；O、P. 裂鳞果（未定种）；
Q~W. 石籽（未定种 1~5）

A–D, *Ferganella* sp.: A and B showing leaves, C and D showing disoriental stomata in lower cuticles;
E. *Elatocladus pinnata*; F. *Cupressinocladus?* sp.; G. *Taxodium?* sp.; H, I. *Elatides ovalis?*; J, K.
*Scarburgia baiyangheensis*; L. *Elatocladus* sp.1; M, N. *Strobilites* spp.1, 2; O, P. *Schizolepis* sp.; Q–W.
*Carpolithus* spp. 1–5

31. 披针型林德勒枝 *Lindlecladus laceolatus*（L. et H.）Harris

32. 费尔干杉（未定种）*Ferganella* sp.

33. 白杨河斯卡布果 *Scarburgia baiyangheensis* Yang et al.

34. 卵形似枞 ? *Elatides ovalis* Heer?

35. 小枞型枝 *Elatocladus minutus* Doludenko

36. 短叶枞型枝 *Elatocladus pinnata* Sun et Zheng

37、38. 枞型枝（未定种 1、2）*Elatocladus* spp. 1，2

39、40. 似果穗（未定种 1、2）*Strobilites* spp. 1，2

**分类不明的化石（Unclassified taxa）及木化石 （Fossil woods）**

41~45. 石籽（未定种 1~5）*Carpolithus* spp. 1–5

46. 木化石（待研究）Fossil wood（in study）

这一植物群总体上显示了中国北方以锥叶蕨 – 拟刺葵 – 银杏（*Coniopteris-Phoenicopsis-Ginkgo*）组合为特征的中侏罗世植物群面貌，属于中国锥叶蕨 – 拟刺葵植物群（*Coniopteris–Phoenicopsis* flora）的晚期部分（周志炎，1995；Sun et al., 2021），主要反映暖温带—温带的温暖潮湿气候，但也存在季节性变化。

### 3.4.2 奥勃鲁契夫银杏

奥勃鲁契夫银杏（*Ginkgo obrutschewi*）是英国古植物学家秀厄德最早在白杨河建立的银杏属分类群。由于该种分布较为广泛，对它的研究无疑得到国内外古植物学家的关注。当年，限于奥勃鲁契夫的有限采集，秀厄德仅报道了 3 件保存并不完整的奥勃鲁契夫银杏化石及它们的零星表皮构造，并只有普通光学显微镜下的观察结果（Seward，1911）。

但近十余年来，本书作者在白杨河该种的模式产地又补充采集了一批新的奥勃鲁契夫银杏化石，详细研究了该种叶化石表皮构造（图 32，D~F），研究取得重要进展（苗雨雁，2006，2017；Sun et al.，2004a，2010，2021）。新的研究显示，奥勃鲁契夫银杏叶表皮构造主要特征是：（1）表皮构造基本上为下气孔式（hypostomatal）；（2）表皮细胞平周壁普遍发育强度不等的角质化丘状突起，垂周壁略直或微弯；（3）气孔器单唇式，保卫细胞内侧强烈唇状加厚，副卫细胞通常也强烈角质化。

　　总的看来，奥勃鲁契夫银杏叶上表皮的普通表皮细胞通常为不规则的短多边形，垂周壁常角质化加厚，呈微波状弯曲。叶下表皮非气孔带由 9~14 列长矩形或长梭形细胞组成；气孔带内的普通表皮细胞常为近等轴状多边形，气孔器单唇型、椭圆形，孔缝通常不定向，保卫细胞在近孔缝处唇状加厚，其上有时还可见自孔缝向外辐射状伸展的角质化加厚纹；副卫细胞 4~5 个，通常强烈角质化，表面通常发育瘤状或团块状角质化加厚；气孔带内的气孔密度约 31~36 个 / 平方毫米。

　　一项很有意义的工作是，前不久俄罗斯青年学者诺索娃等关于奥勃鲁契夫银杏的补充研究工作。诺索娃等补充拍摄了保存在俄罗斯地质博物馆的该种模式标本（No. 29）和地模标本（No. 31），并进行了表皮构造的研究（图 34，J~P），具有重要参考价值（Nosova et al.，2011）。当然，对她提出的该种具有"双面气孔式"的意见，本书作者认为还值得进一步研究（苗雨雁，2017）。此外，孙革还向诺索娃介绍了新疆白杨河和福海的具体地理位置，对俄罗斯学者研究福海沙吉海村加尔煤矿（中心地理位置：86°37'E, 46°40'N）及其相关研究及时提供了帮助；目前，孙革已与诺索娃开始有关中国东北晚白垩世植物群的研究（图 34）。

## 3.5 硫磺沟侏罗纪哺乳动物组合

　　侏罗纪时期的新疆，早期哺乳动物已初见端倪。准噶尔盆地乌鲁木齐以西约 50 千米的硫磺沟是著名的侏罗纪哺乳动物化石产地，这里的晚侏罗世齐古组已发现较丰富的哺乳动物化石，本书称之为"硫磺沟侏罗纪哺乳动物群"。2001~2010 年，本科研队以弗莱茨纳和马丁为领导的课题组在这里首次发现了以准噶尔兽 – 中华艾榴齿兽 – 梯格兽 – 侏掠兽（*Dsungarodon-Sineleutherus-Tegotherium-Nanolestes*）组合为代表的侏罗纪哺乳动物群，该动物群以柱齿兽

**图 34 奥勃鲁契夫银杏化石新研究**

Fig. 34 New studies of fossil *Ginkgo obrutschewi* Seward

A~D. 奥勃鲁契夫银杏化石；E. 奥勃鲁契夫银杏模式标本；F、G. 地模标本；H、I. 秀厄德
研究的表皮构造（1911）；J、K. 新拍的模式及地模标本；L~P. 诺索娃等研究的表皮构造；
Q~V. 作者采集的标本及研究的表皮构造，其中 T 示上表皮，U 和 V 示下表皮及气孔；
W. 孙革在俄罗斯地质博物馆研究白杨河化石（2006）；X. 诺索娃与孙革在中国（2019）；
Y. 苗雨雁在白杨河（2005）

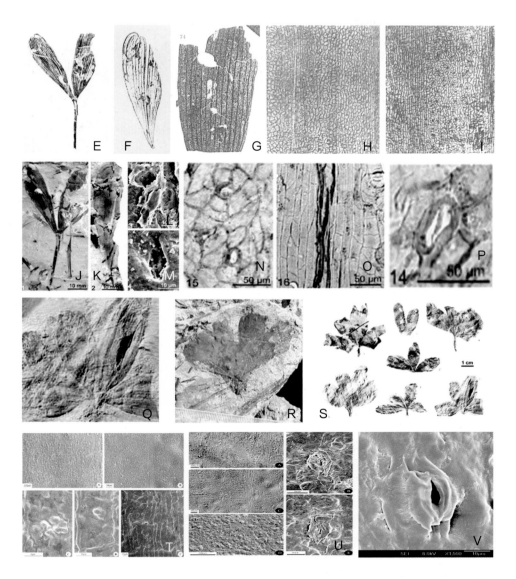

A–D. *Ginkgo obrutschewi*; E. Holotype of *Ginkgo obrutschewi*; F, G. Topotypes; H, I. Cuticles studied by Seward（1911）; J–P. New photos of the holotype and topotype, and the cuticles studied by Nosova et al.; Q–V. The specimens of *G. obrutschewi* newly collected, and cuticles studied by the authors: T showing the upper cuticles, U and V showing the lower cuticles with stomata; W. Sun G. studying fossils from Baiyanghe in VSEGEI（2006）; X. Dr. Nosova and Sun G. in China（2019）; Y. Miao Y. Y. in Baiyanghe（2005）

类动物为主，代表了亚洲晚侏罗世分异度最高的哺乳动物组合（Pfretzschner et al.，2004，2005，Martin et al.，2010b）（图35）。目前已发现的哺乳动物化石主要为牙齿及下颌骨，主要由至少5个分类群组成，分别为：

**柱齿兽类（Docodonta）**

左氏准噶尔兽 *Dsungarodon zuoi* Pfretzschner et Martin，2005

梯格兽（未定种）*Tegotherium* sp.

**小贼兽类（Haramiyida）**

维吾尔中华艾榴齿兽 *Sineleutherus uyguricus* Martin et al.，2010

**基干兽类（Stem-Zatheria）**

麦氏侏掠兽 *Nanolestes mackennai* Martin et al.，2010

**真三尖齿兽类（Eutriconodonta）**

真三尖齿兽类（未定属种）Eutriconodonta indet.

## （1）左氏准噶尔兽 *Dsungarodon zuoi* Pfretzschner et Martin, 2005

准噶尔兽（*Dsungarodon*）是本书作者马丁和弗莱茨纳于硫磺沟晚侏罗世齐古组首建的新属（Pfretzschner et al.，2005），属于柱齿兽类的柱齿兽科（Docodontidae）。2010年，马丁等对该属属征做了修订，主要特征为：其下臼齿具有一个由a-b、a-g和b-g等三条脊围成的大假跟座，下臼齿冠远端具褶，在麦氏沟上方有一额外的沟与后齿骨槽分开，所有这些特征均有别于其他柱齿兽类已知分类群（Martin et al.，2010b）（图35，D）。

左氏准噶尔兽（*D. zuoi*）是准噶尔兽的模式种，种名赠予新疆地矿局第一区调大队前大队长左学义，感谢他对中德新疆地质工作站所作出的贡献。柱齿兽类（Docodonta）最早发现于墨西哥北部及美国南部，此后也发现于中亚地区的哈萨克斯坦及吉尔吉斯斯坦等地，均为晚侏罗世地层，并主要发现的化石均为牙齿。2019年马丁等曾在中国内蒙古东部上侏罗统发现了保存完整个体的柱齿兽类化石——微小柱齿兽（*Microdocodon*），体长仅15厘米，形似现今的鼠类动物（Zhou et al.，2019）（图35，J）。以往报道的在准噶尔盆地五彩湾晚侏罗世石树沟组上部的尖钝齿兽（*Acuodulodon* Hu et al.，

2007）应是准噶尔兽的同异名（junior synonym）。

**（2）梯格兽（未定种）*Tegotherium* sp.**

本分类群属于柱齿兽类的梯格兽科（Tegotheriidae），本属为塔塔里诺夫（Tatarinov）1994 年于蒙古沙尔梯格地区上侏罗统所建。与柱齿兽科不同的是：本属的假下根座由 a–g、a–b、b–e 和 g–e 脊围绕。与塔什库梅尔兽（*Tashkumyrodon*）相比，本属 a–d 脊强而 c–d 脊缺失，而后者 c–d 脊强且 a–d 脊缺失。与克鲁萨特兽（*Krusatodon*）相比，本属下臼齿冠远端缺失额外的脊。此外，本属还具有完整的舌侧齿带。与西伯利亚齿兽（*Sibirotherium*）相比，本属的齿脊 e–g 更强，而远侧齿带（d–dd 脊）要弱得多。此外，本属与克鲁萨特兽和西伯利亚齿兽的不同之处还在于：本属的上臼齿的舌侧齿尖是 2 个（X 和 Y），而不是 3 个（X、Y 和 Z）（图 35，E）。

**（3）维吾尔中华艾榴齿兽 *Sineleutherus uyguricus* Martin, Averianov et Pfretzschner, 2010**

本分类群是作者科研队成员（2010）发现的新属种。中华艾榴齿兽（*Sineleutherus*）属于柱齿兽类的艾榴齿兽科（Eleutherodontidae），其主要特征为：下臼齿呈卵形，有两排在远端周围连续的齿尖列；最大的牙尖位于唇排近中端的下臼齿。与异兽亚纲多瘤齿兽目的不同之处在于保留了宽阔的嚼面轮廓；与该目代表属——艾榴齿兽属（*Eleutherodon*）相比，当前的下臼齿的边缘齿尖更大且更少，并且中央谷盆缺失大量的横向脊（沟）（图 35，F）。

**（4）麦氏侏掠兽 *Nanolestes mackennai* Martin, Averianov et Pfretzschner, 2010**

侏掠兽属（*Nanolestes*）是马丁 2002 年所建，属于基干兽类（stem–Zatheria）的"双掠兽科"（"Amphitheriidae"）。麦氏侏掠兽是马丁等（2010）新建，该新种名是誉予古哺乳动物学家麦克纳（McKenna M. C., 1930~2008），以感谢她对中生代哺乳动物研究作出的贡献。本属的修订特征主要为：下臼齿三角座较短（呈 50° 夹角），下后尖与下原尖在一条横线上，下后尖不向后移（Martin et al., 2010b）（图 35，G）。

哺乳动物化石点

**图 35　晚侏罗世硫磺沟哺乳动物群**

Fig. 35　Late Jurassic mammal fauna from Liuhuanggou of Junggar, Xinjiang

A. 硫磺沟；B. 化石产地：硫磺沟，晚侏罗世齐古组；C. 弗莱茨纳（右）与马丁在硫磺沟野外（2005）；D. 左氏准噶尔兽；E. 梯格兽（未定种）；F. 维吾尔中华艾榴齿兽；G. 麦氏侏掠兽；H. 真三尖齿兽目（未定属种）；I. 德国出版本成果的杂志；J. 柱齿兽生态参考图片（微小柱齿兽，马丁等在辽宁晚侏罗世地层发现，2019）；K~N. 在硫磺沟野外挑选化石样品、水洗与搬运，L 右为孙跃武教授（2005）（D 和 E 据 Martin et al., 2010b）

A. Valley of Liuhuanggou; B. Fossil site: Upper Jurassic Qigu Formation in Liuhuanggou; C. Prof. Pfretzschner H.–U.（right）and Prof. Martin T. in field work at Liuhuanggou, 2005; D. *Dsungarodon zuoi*; E. *Tegotherium* sp.; F. *Sineleutherus uyguricus*; G. *Nanolestes mackennai*; H. *Eutriconodonta indet.*; I. The journal publishing the study results, 2010; J. A referring photo showing Microdoncodon found by Martin T. et al., from Upper Jurassic of Liaoning, China（2019）; K–N. Fossil collecting, washing and transporting, L right showing Prof. Sun Y. W.（2005）（D and E after Martin et al., 2010b）

#### （5）真三尖齿兽目（未定属种）Eutriconodonta indet

本分类群属于真三尖齿兽目（Eutriconodonta），但属种级详细分类尚待进一步研究。本分类群主要特征为：牙齿有一个大的主尖a，略微向远端弯曲，其顶点缺失；一个相对较大、约1/2高度的远端主尖c，很尖，顶端朝向背侧；一个小但明显的、更近远端主尖d，与齿带状体分离。牙冠远端很尖，没有明显的远端齿带；舌侧和唇侧齿带延伸到牙齿的尖端，前者比后者长约2倍。在两侧，齿带仅局限于牙冠的后部，不会向近中延伸超过c尖。从咬合面看，牙冠在后侧略不对称，舌侧更扩张，但前侧更对称。牙釉质表面雕刻有非常精细的条纹。牙齿双根，根间有很大的空间，但根未完整保留（图35，H）。

硫磺沟哺乳动物群在新疆准噶尔盆地晚侏罗世齐古组的首次发现，展示了晚侏罗世时期的早期哺乳动物在亚洲地区的最高分异度，也表现出它们与蒙古沙尔梯格晚侏罗世哺乳动物组合（产柱齿兽类 *Tegotherium gubini* Tatarinov，1994）和中国四川世龙寨晚侏罗世沙溪庙组（上部）哺乳动物组合（产蜀兽 *Shuotherium*）的相似性。当然，蜀兽（*Shuotherium*）的时代也很可能属于中侏罗世，因为在英国柯特林顿，具可靠时代证据的巴统期地层也发现过该属化石。此外，硫磺沟哺乳动物群与英国大理岩林（Forest Marble）巴统期的哺乳动物组合也有一定的相似性。尽管硫磺沟哺乳动物组合在缺少碟齿兽类（Dryolestidans）和多瘤齿兽类（Multituberculates）等方面反映出与上述英国两个中侏罗世哺乳动物组合存在一定差异，但上述相似性似暗示，硫磺沟哺乳动物组合可能也保留了一些中侏罗世哺乳动物的原始特征（Martin et al.，2010b）。

## 3.6 鄯善侏罗纪恐龙足迹群

鄯善恐龙足迹群发现于七克台南戈壁滩，中侏罗世三间房组灰绿 – 黄绿色砂岩夹紫色砂岩及粉砂岩中。已发现的恐龙足印共155个，呈密集分布，不规则排列，分布范围长超过100米（图36）。已发现的恐龙足迹均为三趾型，每个脚趾前具一尖爪，均属肉食性的兽脚类恐龙。每个足印均显示了 3 个大而向前伸出的脚趾并具有大而尖的爪，至少包括两种不同的形态型：一类（形态型 A）足印较大，有相对宽大的足垫，趾长可达 33 厘米的脚印，可能由大型肉食龙类形成；另一类（形态型 B）足印为较小而纤细型，可能由轻巧的虚骨龙类形成。

具体分类描述如下（Wings et al.，2007）：

**形态型 A**：足迹长度大于宽度，呈近三角形。宽度为 17.5~38.2 厘米。足跟印迹较为明显。第Ⅲ趾最长，长度为 18.3~33.6 厘米；第Ⅱ趾与第Ⅳ趾的长度约为第Ⅲ趾长度的 75%，第Ⅱ趾稍长于第Ⅳ趾。第Ⅱ趾与第Ⅲ趾之间的平均角度为 37°，第Ⅲ趾与第Ⅳ趾之间的平均角度为 40°，足迹整体角度为 77°。第Ⅱ趾与第Ⅲ趾尖端向内侧弯曲，并有明显的"V"形爪印（图 37，A）。

**形态型 B**：足迹较长，纤细。宽度为 12.2~33.3 厘米。保存有足跟印迹但不明显。第Ⅲ趾最长，长度为 11.4~29.2 厘米；第Ⅱ趾与第Ⅳ趾长度近等，约为第Ⅲ趾长度的 70%，第Ⅱ趾稍长于第Ⅳ趾。第Ⅱ趾与第Ⅲ趾之间的平均角度为 37°，第Ⅲ趾与第Ⅳ趾之间的平均角度为 40°，足迹整体角度为 73°，第Ⅱ、第Ⅲ、第Ⅳ趾之间的平均角度近等。所有趾尖端均有明显的"V"形爪印，但比形态型 A 的爪印稍小（图 37，B）。

中国专家邢立达等重新研究了这批足迹（Xing et al.，2014），认为鄯善恐龙足迹群是中国西北新疆首次发现的恐龙足迹记录，有些足迹显示了明显的趾垫、足跟及罕见的跖骨末端印痕形态，对研究上述恐龙足迹群的古生态具有重要意义。但邢立达等认为，上述两种不同的足迹类型在外形上出现的差异属于由于软而湿滑的基底造成的额外形态变化，或由不同年龄段的同类造迹者所留，据此将其均归入石炭张北足迹（*Changpeipus carbonicus*）。其中，较大型的足迹有相对宽的趾垫，他们认为与张北足迹及巨齿龙足迹（*Megalosauripus*）较为相似；另一类相对更细小的足迹与似鹬龙足迹（*Grallator*）、实雷龙足迹（*Eubrontes*）及安琪龙足迹（*Anchisauripus*）较为相似。

本书作者认为，鄯善恐龙足迹群中的小型足迹与大型足迹还是存在明显

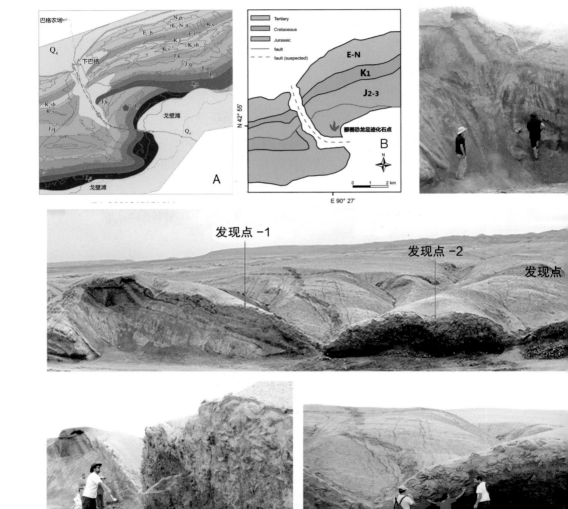

**图 36　鄯善侏罗纪恐龙足迹群**

Fig. 36　The fossil dinosa ur tracks in Shanshan

A、B. 鄯善恐龙足迹群地质地理分布；C. 发现的第 1~3 恐龙足迹群出露点；
D~F. 青年专家在剖面工作；G、H. 恐龙足印；I. 专家在化石点（2008）

A, B. Sketchs of the geography and stratigraphy of the dinosaur track site in Shanshan；C. No. 1–3 exposive sites of dinosaur tracks；D–F. Young scientists working in the outcrops（2007）；G, H. Dinosaur tracks；I. Experts at the dinosaur track site（2008）

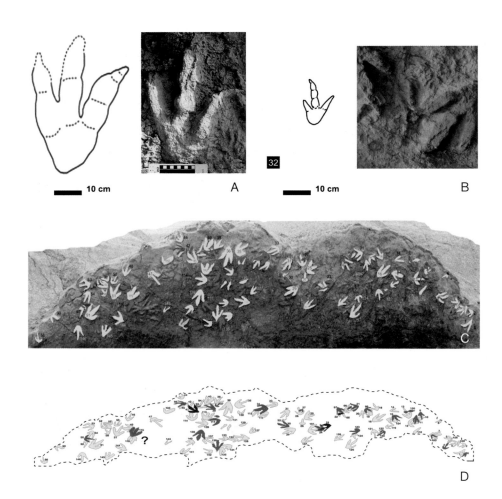

**图 37 鄯善侏罗纪恐龙足迹类型**

Fig. 37 Types of Jurassic dinosaur tracks in Shanshan

A. 恐龙足迹类型 A；B. 恐龙足迹类型 B（据 Wings et al., 2007）；C、D. 第 2 出露点鄯善恐龙足迹群平面分布图

A. Morphology A of dinosaur tracks; B. Morphology B of dinosaur tracks（Wings et al., 2007）；C，D. Plan of dinosaur tracks from exposive site No. 2

区别的，目前暂时还是以划分出两种类型为宜，有待进一步研究。

　　此外，本科研队在恐龙足迹所在的三间房组及其上部层位，还发现无脊椎动物遗迹化石、蜥脚类恐龙骨骼及硅化木等。综合分析鄯善侏罗纪恐龙足迹，其代表的恐龙动物群所反映的古生态环境是湖边沼泽相，即中侏罗世时期，新疆鄯善的恐龙曾生活在吐鲁番地区湖边的沼泽一带，距离附近的森林不远，当时的气候温暖而潮湿。邢立达等（2014）认为，丰富的无脊椎动物遗迹化石支持该化石点的沉积环境是一个逐步扩大和深化的湖泊。此地的无脊椎动物遗迹化石洛克迹（*Lockeia*）应归入福尔斯迹（*Fuersichnus*），这类遗迹被归于泥食性昆虫幼虫或其他无脊椎动物所留。福尔斯迹指示的是在泥泞的河漫滩或湖泊边缘居住和（或）觅食（Xing et al.，2014）。

# 3.7　鄯善新疆巨龙

　　新疆巨龙（*Xinjiangotitan*）产于鄯善七克台以南的晚侏罗世齐古组，该组由杂色砂岩、粉砂岩及细砾岩，局部夹泥岩组成。该恐龙化石是 2011~2013 年由本科研队在七克台原地化石埋藏点采得，2013 年由科研队年轻的恐龙专家吴文昊等命名，模式种为鄯善新疆巨龙（*Xinjiangotitan shanshanensis* Wu et al.，2013），参加本研究的包括维恩斯、周长付、关谷透（Sekiya T.，日本青年专家）及董枝明等（吴文昊等，2013）。截至 2013 年 6 月，本科研队已采得鄯善新疆巨龙的部分颈椎（3 节）、完整的背椎（12 节）、荐椎（5 节）、腰带（坐骨、肠骨、耻骨），以及左侧股骨、胫骨及腓骨等，股骨长约 1.6 米。推测该恐龙，体长约 30 米，是中国迄今保存最好、形体最大的侏罗纪恐龙化石（图 38）。

　　吴文昊等（2013）对鄯善新疆巨龙做了分类支序分析研究，认为该分类群在系统分类上与已知的马门溪龙属（*Mamenchisaurus*）互成姊妹群，由一个确定的共近裔性状支持：后部背椎呈后凹型。据此认为，两者同属于蜥臀目的蜥脚亚目（Sauropoda）马门溪龙科（Mamenchisauridae Young et Chao，

J

C

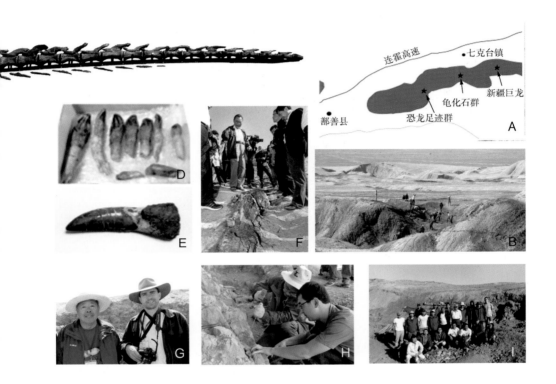

**图 38 鄯善新疆巨龙**

**Fig. 38 _Xinjiangotitan shanshanensis_ Wu et al., 2013**

A. 鄯善新疆巨龙化石点位置示意图；B. 恐龙化石发掘；C~E. 已发掘的部分恐龙化石，其中 D 为新疆巨龙牙齿，E 为兽脚类恐龙牙齿；F~I. 野外发掘工作；J. 新疆巨龙复原骨架（原照片由鄯善县国土资源局提供，2016）

A. Sketch illustration on the geographic position of _Xinjiangotitan_; B. Excavation view; C. Skeleton of _Xinjiangotitan_ in situ; D. teeth of _Xinjiangotitan_; E. Theropod tooth; F–I. Field work during excavation; J. Reconstruction of _Xinjiangotitan_（photo provided by Bureau of Land & Resources, Shanshan County, 2016）

1972）（图 39）。

2016~2018 年，中国的恐龙学专家李大庆率领团队对鄯善新疆巨龙进行了后续发掘工作，取得重要进展。除发现恐龙的部分头骨外，还发现了完整的颈椎（18 节）及 39 节尾椎等，使鄯善新疆巨龙以一个基本完整的恐龙骨架展现出来，2018 年又进一步详细描述，提高了研究程度（Zhang et al., 2018）（图40）。根据 Zhang 等（2018）的最新描述，目前，鄯善新疆巨龙（模式标本）已发现的组成部分包括：部分头骨、18 节颈椎、12 节背椎、5 节荐椎、39 节尾椎；部分颈肋和背肋，以及 18 个脉弧；左侧肠骨、两侧耻骨、两侧坐骨，左侧的股骨、胫骨及腓骨；左侧距骨；一部分左足等（Zhang et al., 2018）。近完整的标本包括完整且几乎完全相关节的颈椎，颈椎长至少 14.9 米，由 18 节椎体组成；最后一节颈椎与第一节背椎相关节；所有颈椎椎体侧面发育纵向气窝（longitudinal pneumatic fossa）；第 3~16 颈椎的纵向气窝被一斜向嵴分隔成前、后两部分。神经棘较低，在第 3~14 节颈椎中为前后延长、近水平；在第 16、第 17 节颈椎中变短；除第 18 节颈椎，神经棘不分叉。背椎共 12 节，总长 2.63 米；荐椎共 5 节，椎体完全愈合，长度近等；荐椎腹侧相对背椎变窄，椎体腹侧不发育嵴或沟；与其他蜥脚类恐龙不同的是，它第 1、第 2 节荐椎侧面同时被第 2 荐肋连接，第 1 荐肋不参与构成荐椎轭（sacriocostal yoke）；耻骨骨干很平，相比其他马门溪龙科类群，新疆巨龙的栖肌突发育更加明显，且向下延伸；左侧股骨完整保存，非常粗壮，长 0.9 米，远端最大宽与股骨长度比为 0.33，近端前后扩张，其前后宽度为股骨长度的 18%。

前不久，李大庆团队的研究（Zhang et al., 2018）认为：鄯善新疆巨龙属真蜥脚类（Eusauropoda）基干类群，具有马门溪龙科恐龙的典型特征，如背椎 12 节，荐椎 5 节，荐前椎后凹型，椎体内具蜂窝构造，颈部长、颈椎椎体延长、颈肋长，前部背椎与后部颈椎神经棘分叉，前部尾椎前凹型等（欧阳辉，叶勇，2002）。但鄯善新疆巨龙的耻骨发育明显的栖肌突、较短的后肢以及股骨第 4 转子位于股骨后侧内缘等特征，与更进步的新蜥脚类（Neosauropoda）中的梁龙超科（Diplodocoidea）相似，而与马门溪龙科不同（图40）。

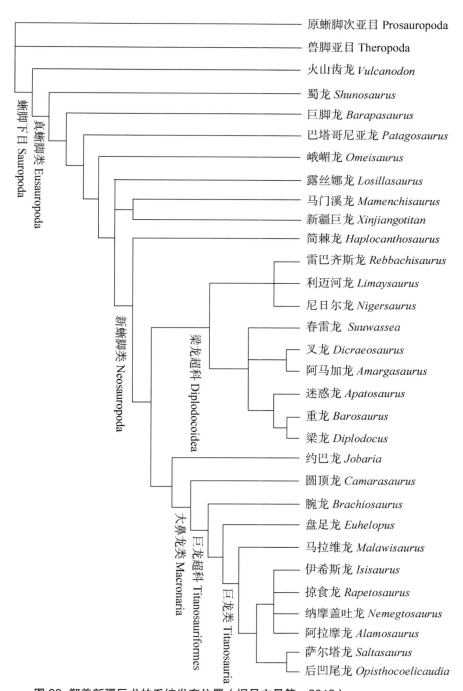

图 39　鄯善新疆巨龙的系统发育位置（据吴文昊等，2013）

Fig. 39　Phylogenetics of *Xinjiangotitan shanshanensis* (after Wu et al., 2013)

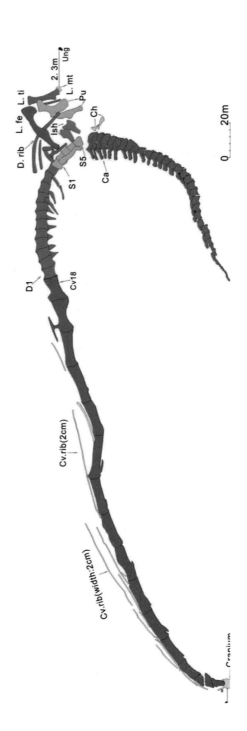

图 40 鄯善新疆巨龙埋藏情况图（据 Zhang et al.，2018）

Fig. 40 Illustration of *Xinjiangotitan* in bury (after Zhang et al., 2018)

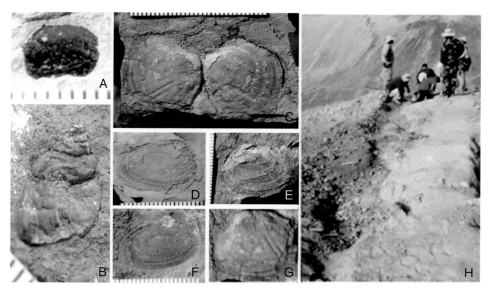

**图 41　新疆巨龙附近的恐龙化石及下伏地层的化石的发现**

**Fig. 41　Discoveries of dinosaurs associated and the other fossils in underlying beds**

A. 鱼鳞化石；B. 腹足类化石；C~G. 瓣鳃类化石；H. 东 1 点恐龙化石发现（B~G 为下伏地层所产）

A. Fossil fish scale; B. Gastropod; C–G. Bivalves; H. new discovery of dinosaur vertebrae at East Point No. 1（B–G from the underlying strata）

　　除新疆巨龙外，本科研队还在新疆巨龙附近同层位发现了兽脚类牙齿化石；在新疆巨龙化石点附近（东 1 点，中心地理位置：42°57'47.6"N，90°34'32.5"E）发现 8 块关联的颈椎，长约 16 米（图 41，H）。

　　有关齐古组的地质时代，2012 年地科院王思恩等曾在该组获 164.6 ± 1.4 Ma测年值（王思恩等，2012），因此原认为是晚侏罗世早期。

　　此外，在产新疆巨龙的齐古组下伏的七克台组，本书作者还发现鱼鳞、腹足类及瓣鳃类等化石。瓣鳃类化石经沙金庚鉴定，包括有 *Kajia ovalis*（Martinson），以及 *Kijia kweizhouensis*（Grabau）等（图 41，A~G）；上述瓣鳃类化石以往曾见于新疆塔里木盆地库车地区的恰克马克组和克孜勒努尔组，时代为中侏罗世。这些瓣鳃类化石为产新疆巨龙的层位齐古组的时代属于晚侏罗世早期，似提供了旁证。

# 3.8 沙尔湖中侏罗世森林

沙尔湖煤田位于新疆鄯善县境内、吐哈盆地南缘中段，西距鄯善县城约 130 千米，东距哈密市约 160 千米，中心地理位置的坐标为 91° 20′ E，42° 35′ N。

沙尔湖中侏罗世西山窑组主要由砂岩、粉砂岩、泥岩、炭质泥岩等组成，富含巨厚煤层，并产丰富的植物（包括孢粉）化石。新疆地矿局第一地质大队（1989）最早完成了对沙尔湖煤田煤炭资源的远景调查工作，根据报告所示，中侏罗统西山窑组含煤地层层序如下（董曼等，2011）（图 42）。

**上覆地层: 中侏罗统头屯河组**            **厚度（米）**
砖红色泥岩、泥质粉砂岩、粉砂岩、细粒砂岩
————————整 合————————

**西山窑组**
4. 上部为灰绿色泥岩、粉砂岩夹碳质泥岩和煤层
    下部灰绿色粉砂岩、中细粒砂岩等 ———————— 143.59
3. 上部为深灰色泥岩、砂岩、碳质泥岩和煤层
    下部为灰色泥质粉砂岩夹碳质泥岩和煤层 ———————— 93.92
2. 深灰色和浅灰色泥质粉砂岩、碳质泥岩和煤层
      夹粉砂质泥岩 ———————————————— 85.36
1. 灰色粉砂岩、泥质粉砂岩、砂岩、碳质砂岩和
    煤层。产大量植物化石 ———————————— 185.59
————————不 整 合————————

**下伏地层: 下二叠统 阿奇克布拉克组**
      灰绿色玄武岩、安山岩，凝灰质砂岩和角砾岩等

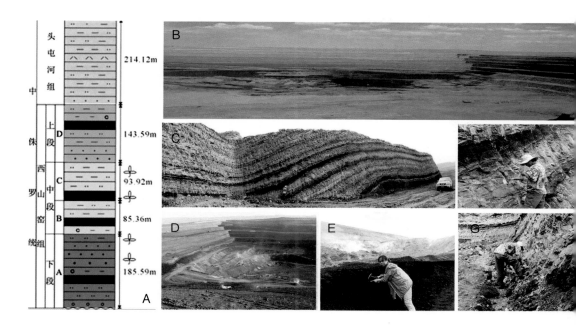

**图 42　新疆沙尔湖中侏罗世植物群**

Fig. 42　Middle Jurassic Shaerhu flora of Xinjiang

A. 地层柱状图；B~D. 沙尔湖煤田及含煤地层；E~G. 野外工作（E 为董曼，F、G 为杨涛）

A. Stratigraphic column; B~D. Outlines of the Shaerhu Coal–Mine and its coal–bearing strata; E~G. Field working in Shaerhu: E. Dr. Dong M., F, G. Dr. Yang T.

## 3.8.1　植物大化石

　　本科研队董曼等在煤田中部偏西北方向的西山窑组中、下段的煤层夹矸粉砂岩或泥质粉砂岩之中，采集植物大化石 227 件，孢粉样品 35 件，另有一部分孢粉样品采自煤层。现已发现的植物大化石至少有 18 属 20 种。该植物群以苏铁类及蕨类占优势为特征，银杏类、茨康类及松柏类等也占有一定比例。其中，有节类 2 属 2 种（占 10.0%），真蕨类 4 属 6 种（占 30.0%），苏铁类及本内苏铁类 4 属 6 种（占 30.0%），银杏类及茨康类共 4 属 5 种（占 20.0%），松柏类 2 属 2 种（占 10.0%）（董曼等，2011）。大化石植物群主要组成分子如下（图 43~ 图 45）：

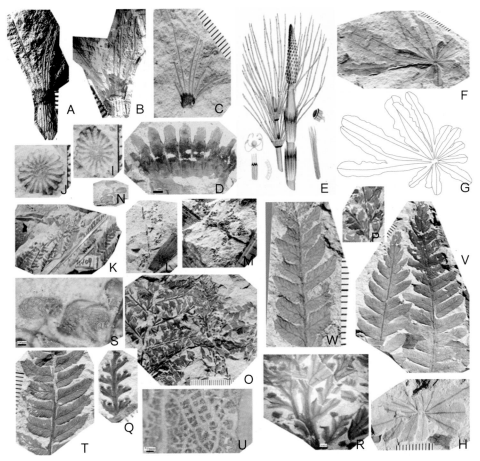

**图 43 沙尔湖植物群的蕨类**

Fig. 43 The ferns of Shaerhu flora

A~E、I、J. 沙尔湖似木贼，D 示叶鞘，E 为复原图，I、J 示横膈膜；F~H. 辛浦生拟轮叶，G 为 F 的素描图；K~N. 拟紫萁（未定种），N 为孢子囊群的放大；O~S. 膜蕨型锥叶蕨，Q~S 为生殖叶，S 示孢子囊群放大；T、U. 狄阿姆拉法尔蕨，U 示叶脉结构；V. 纤柔枝脉蕨（比较种）；W. 怀特枝脉蕨

A~E、I、J. *Equisetites shaerhuensis*, D showing leaf sheath, E showing restoration map, I and J represent the diaphragm; F~H. *Annulariopsis simpsoni*, G is the sketch map of F; K~N. *Osmundopsis* sp., N is the enlargement of the sporangia; O~S. *Coniopteris hymenophylloides*, Q~S is the reproductive leaf, S showing the enlargement of the sporangia; T, U. *Raphaelia diamensis*, U showing vein structure; V. *Cladophlebis* cf. *delicatula*; W. *Cladophlebis whitbiensis*

**有节类 Equisetales**

1. *Equisetites shaerhuensis* Dong 沙尔湖似木贼

2. *Annulariopsis simpsoni*（Phill.）Harris 辛浦生拟轮叶

**真蕨类 Ficinae**

3. *Coniopteris hymenophylloides* Brongniart 膜蕨型锥叶蕨

4. *Osmundopsis sturi*（Raciborski）Harris 司图尔拟紫萁

5. *Cladophlebis whitbiensis* Harris 怀特枝脉蕨

6. *Cladophlebis* cf. *delicatula* Yabe et Oishi 纤柔枝脉蕨（比较种）

7. *Raphaelia diamensis* Seward 狄阿姆拉法尔蕨

**本内苏铁类 Bennetiatales**

8. *Pterophyllum propinquum* Goeppert 紧密侧羽叶

9. *Nilssoniopteris* sp. 尼尔桑带羽叶（未定种）

**苏铁类 Cycadales**

10. *Nilssonia* cf. *acuminata* Presl 渐尖尼尔桑（比较种）

11. *Nilssonia* cf. *tenuinervis* Nathorst 密脉尼尔桑（比较种）

12. *Ctenis* cf. *kaneharai* Yokoyama 金原篦羽叶（比较种）

**银杏类 Ginkgoales**

13. *Ginkgo digitata*（Brongn.）Heer 指状银杏

14. *Ginkgo obrutschewi* Seward 奥勃鲁契夫银杏

15. *Ginkgoites* cf. *lepidus*（Heer）Florin 长叶似银杏（比较种）

**茨康类 Czekanowskiales**

16. *Phoenicopsis angustifolia* Heer 狭叶拟刺葵

17. *Czekanowskia* sp. 茨康叶（未定种）

**松柏类 Coniferales**

18. *Lindleycladus lanceolatus*（L. et H.）Harris 披针型林德勒枝

19. *Strobilites* sp. 似果穗（未定种）

**分类不明的分类群 Unclassified taxa**

20. *Carpolithus* sp. 石籽（未定种）

蕨类植物中，有节类包括沙尔湖似木贼及辛浦生拟轮叶等。沙尔湖似木贼是董曼建立的新种，其叶鞘等颇具特征，与该属已知种不同（董曼，2012）。为详细了解该种特征，本书特进一步总结叙述如下。

**沙尔湖似木贼 *Equisetites shaerhuensis* Dong**（图 44），（据董曼，2012，页 28，图版 1，图 1~3，6~8；图版 2，图 1，图 2）

特征：具叶茎宽线性，分节与节间，节间长 1 厘米以上，宽 3.5~4 毫米，纵脊与纵沟相间排列；节部具细长的叶，约 17 枚，长 1.5~2.1 厘米，宽 0.5 毫米，叶披针形，中脉明显；叶底部连合成叶鞘，每枚叶鞘长 3~4 毫米，宽 0.6~0.7 毫米，在 1/2~1/3 处分裂，叶鞘基部见有极细的纵纹，表皮细胞呈伸长的不规则四边形，长 33~166 微米，宽 20~50 微米；关节盘椭圆形，外环长茎约 1.1 毫米，内环长茎约 0.3 毫米；辐射线 17 条，自内环射出，具肋状突起。这一新发现的种级分类群以较独特的叶鞘、叶与关节盘（横膈膜）数目对应等特征区别于该属已知种。

产地及层位：新疆沙尔湖，中侏罗世西山窑组。

真蕨类除膜蕨型锥叶蕨（*Coniopteris hymenophylloides*）外，还发现斯图尔拟紫萁（*Osmundopsis sturi*），推测该种可能是 *Cladophlebis* 的生殖羽片；此外，还发现怀特枝脉蕨（*Cladophlebis whitbiensis*）、纤柔枝脉蕨（比较种）（*Cladophlebis* cf. *delicatula*），以及狄阿姆拉法尔蕨（*Raphaelia diamensis*）等。膜蕨型锥叶蕨很丰富，该种实小羽片退缩成瘦狭的柄状，圆形孢子囊群生于楔羊齿型叶脉的顶端，孢子囊托呈半椭圆形，最宽处直径 0.73~1.10 毫米，并向轴收缩变狭；可见到残留的椭圆形孢子囊，大小约 227 微米 × 172 微米，环带宽约 33 微米（图 43，O~S）。

苏铁类及本内苏铁类较丰富，本内苏铁类包括紧密侧羽叶（*Pterophyllum propinquum*）、较小异羽叶（比较种）（*Anomozamites* cf. *minor*）、尼尔桑带羽叶（未定种）（*Nilssoniopteris* sp.）及带羊齿（未定种）（*Taeniopteris* sp.）等。苏铁类包括渐尖尼尔桑（比较种）（*Nilssonia* cf. *acuminata*）、密脉尼尔桑（比较种）（*Nilssonia* cf. *tenuinervis*）及金原篦羽叶（*Ctenis* cf. *kaneharai*）等。银杏类和茨康类也较丰富，银杏类主要有指状银杏（*Ginkgo digitata*）、奥勃鲁契夫银杏（*G. obrutschewi*）及长叶似银杏（*Ginkgoites* cf. *lepidus*）等；茨康类主要有狭叶拟刺葵（*Phoenicopsis angustifolia*）及茨康叶（未定种）（*Czekanowskia* sp.）等。松柏类发现的不多，主要有披针型林德勒枝（*Lindleycladus lanceolatus*）等（图 45）。

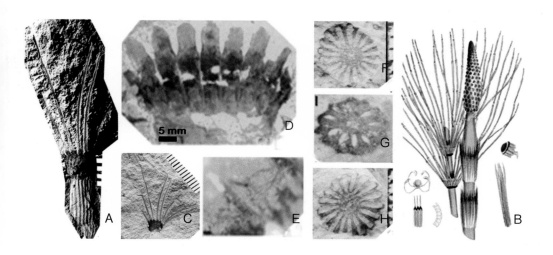

**图 44　沙尔湖似木贼**

Fig. 44 ***Equisetites shaerhuensis*** Dong

A、C. 茎叶；B. 设想复原图（据董曼，2012）；D、E. 叶鞘，其中 E 示叶鞘表皮细胞；F~H. 关节盘

A–C. leafy stems；B. suggested reconstruction（after Dong, 2012）；D, E. sheaths: E showing the epidermal cells of the sheath; F–H. diaphragms

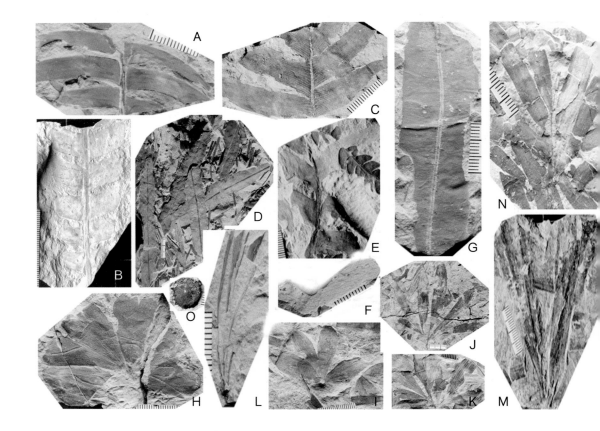

**图 45 沙尔湖植物群的裸子植物**

Fig. 45 Gymnosperms in Shaerhu flora

A. 紧密侧羽叶；B. 较小异羽叶（比较种）；C. 金原篦羽叶（比较种）；D. 带羊齿（未定种）；E、F. 篦羽叶（未定种）；G. 密脉尼尔桑（比较种）；H. 指状银杏；I. 奥勃鲁契夫银杏；J、K. 长叶似银杏（比较种）；L、M. 狭叶拟刺葵；N. 披针型林德勒枝；O. 石籽（未定种）

A. *Pterophyllum propinquum*; B. *Anomozamites* cf. *minor* ; C. *Ctenis* cf. *kaneharai*; D. *Taeniopteris* sp.; E, F. *Ctenis* sp.; G. *Nilssonia* cf. *tenuinervis*; H. *Ginkgo digitata*; I. *Ginkgo obrutschewi*; J, K. *Ginkgoites* cf. *lepidus*; L, M. *Phoenicopsis angustifolia*; N. *Lindleycladus lanceolatus*; O. *Carpolithis* sp.

### 3.8.2 孢粉植物群

　　除植物大化石外，本植物群还发现大量孢粉化石，至少有 34 属 49 种。其中，蕨类植物孢子约占 43%，裸子植物花粉约占 57%。孢子化石以新叉瘤孢－石松孢－阿尔索菲孢－巴洛三角孢（*Neoraistrickia-Lycopodiumsporites-Alsophilidites-Deltoidospora*）组合为代表，花粉以苏铁粉－单束松粉－罗汉松粉（*Cycadopites-Abietineaepollenites-Podocarpidites*）组合为代表。孢粉化石总体显示了明显的中侏罗世孢粉植物群的面貌。主要孢粉组成分子如下（图 46~图 48）：

**孢子 Spores**

1. *Neoraistrickia gristhorpensis*（Couper）Tralau，1968 格里斯索普新叉瘤孢

2. *Alsophilidites arcuatus*（Bolkh.）Xu et Zhang，1980 弓形阿尔索菲孢

3. *Alsophilidites* sp. 阿尔索菲孢（未定种）

4. *Lycopodiumsporites austroclavatidites*（Cookson）Potonie，1956 南方拟棒石松孢

5. *L. subrotundum* （Kara–Mursa）Pocock，1970 近圆石松孢

6. *Lycopodiumsporites* sp. 石松孢（未定种）

7. *Leiotriletes* cf. *adnatoides* Potonie et Kremp，1955 类贴生光面三缝孢（比较种）

8. *Deltoidospora balowensis*（Doring）Zhang，1978 巴洛三角孢

9. *Todisporites major* Couper，1958 大托第蕨孢

10. *Toroisporis*（*Divitoroisporis*）*granularis* Pu et Wu，1985 粒纹具唇孢

11. *Gleicheniidites senonicus* Ross，1949 赛诺里白孢

12. *Gleicheniidites* sp. 里白孢（未定种）

13. *Polypodiisporites* sp. 平瘤水龙骨孢（未定种）

14. *Cyathidites minor* Couper，1953 小桫椤孢

15. *Cyathidites* cf. *concavus*（Bolkh.，1953）Dettmann，1963 凹边桫椤孢（比较种）

16. *Biretisporites* sp. 伯莱梯孢（未定种）

17. *Cibotiumspora corniger*（Bolkh.）Zhang W. P.，1984 具角金毛狗孢

18. *Verrucosisporites granatus*（Bolkh.）Gao et Zhai，1976 多粒圆形块瘤孢

19. *Hsuisporites rugatus* Zhang，1965 皱纹徐氏孢

## 花粉 Pollen

20. *Alisporites minutisaccus* Clarke，1965 小囊阿里粉

21. *Concentrisporites pseudosulcatus*（Briche, Danze, Corsin et Laveine）
Pocock，1970 假沟同心粉

22. *Chasmatosporites minor* Nilsson，1958 较小广口粉

23. *Bennettiteaepollenites Lucifer*（Thierg.）Potonie，1958 鲜明拟本内苏铁粉

24. *Cycadopites adjectus*（De Jersey）De Jersey，1964 具唇苏铁粉

25. *Cycadopites carpentieri*（Delc. et Sprum.）Singh，1964 卡城苏铁粉

26. *Cycadopites* sp. 苏铁粉（未定种）

27. *Inaperturopollenites* sp. 无口器粉（未定种）

28. *Abietineaepollenites minimus* Couper，1958 小单束松粉

29. *Abietineaepollenites* sp. 单束松粉（未定种）

30. *Podocarpidites minisculus* Singh，1964 小罗汉松粉

31. *Podocarpidites canadensis* Pocock，1962 加拿大罗汉松粉

32. *Podocarpidites multisimus*（Bolkh.）Pocock，1962 多凹罗汉松粉

33. *Podocarpidites multicinus*（Bolkh.）Pocock，1970 多分罗汉松粉

34. *Pseudopicea variabiliformis*（Mal.）Bolkh.，1956 多变假云杉粉

35. *Pinuspollenites enodatus*（Bolkh.）Li，1984 光滑双束松粉

36. *Protopinus subluteus* Bolkh.，1956 浅黄原始松粉

37. *Monosulcites minimus* Cookson，1947 小单远极沟粉

38. *Quadraeculina limbata* Maljavkina，1949 真边四字粉

39. *Erlianpollis minisculus* Zhao，1987 较小二连粉

40. *Erlianpollis eminulus* Zhao，1987 微突二连粉

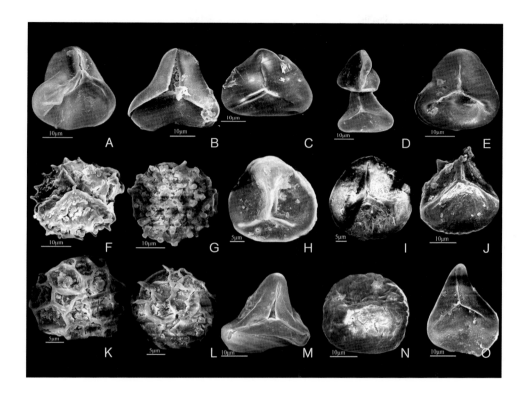

**图 46 沙尔湖植物群孢子化石**

Fig. 46 Spores of Shaerhu flora

A、B. 阿尔索菲孢（未定种）；C~E. 小沙椤孢；F、G. 格里斯索普新叉瘤孢；H、I. 伯莱梯孢（未定种）；J. 具唇孢（未定种）；K、L. 石松孢（未定种）；M. 里白孢（未定种）；N. 平瘤水龙骨孢（未定种）；O. 三角孢（未定种）（据董曼，2012）

A, B. *Alsophilidites* sp.; C–E. *Cyathidites minor*; F, G. *Neoraistrickia gristhorpensis*; H, I. *Biretisporites* sp.; J. *Toroisporis* sp.; K, L. *Lycopodiumsporites* sp.; M. *Gleicheniidites* sp.; N. *Polypodiisporites* sp.; O. *Deltoidospora* sp.（after Dong, 2012）

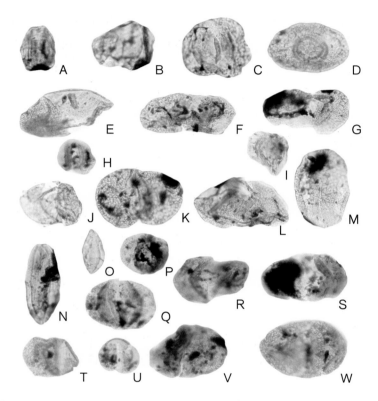

**图 47 中侏罗世沙尔湖植物群的花粉化石（1）**

Fig. 47 Pollen of Middle Jurassic Shaerhu flora (1)

A. 真边四字粉；B. 较小二连粉；C、D. 单束松粉（未定种）；E. 微突二连粉；F. 兰德假瓦
契杉粉；G. 相同云杉粉；H. 小罗汉松粉；I、P、Q、U. 小单束松粉；J. 相同云杉粉；K. 多
凹罗汉松粉；L. 多分罗汉松粉；M. 紧接蝶囊粉；N. 鲜明拟本内苏铁粉；O. 小单远极沟粉；
R. 多变假云杉粉；S、T 光滑双束松粉；V、W. 浅黄原始松粉

A. *Quadraeculina limbata*; B. *Erlianpollis minisculus*; C, D. *Abietineaepollenites* sp.; E. *Erlianpollis eminulus*; F. *Pseudowalchia landesii*; G. *Piceaepollenites omoriciformis*; H. *Podocarpidites minisculus*; I, P, Q, and U. *Abietineaepollenites minimus*; J. *Piceaepollenites omoriciformis*; K. *Podocarpidites multisimus*; L. *Podocarpidites multicinus*; M. *Platysaccus proximus*; N. *Bennettiteaepollenites lucifer*; O. *Monosulcites minimus*; R. *Pseudopicea variabiliformis*; S，T. *Pinuspollenites enodatus*; V，W. *Protopinus subluteus*

41. *Pseudowalchia landesii* Pocock，1970 兰德假瓦契杉粉

42. *Piceaepollenites omoriciformis*（Bolkh.）Xu et Zhang，1980 相同云杉粉

43. *Platysaccus proximus*（Bolkh.）Song，2000 紧接蝶囊粉

44. *Quadraeculina canadensis*（Pocock）Zhang，1978 加拿大四字粉

45. *Rugubivesiculites* sp. 皱体双囊粉（未定种）

46. *Cedripites* sp. 雪松粉（未定种）

47. *Protoconiferus* sp. 原始松柏粉（未定种）

　　总之，沙尔湖植物群以本内苏铁类及苏铁类、真蕨类等占优势为特征，有节类、银杏类、茨康类及松柏类等也占有一定比例，具有典型的中侏罗世早期的色彩，时代可能为中侏罗世早期［即阿伦期 – 巴柔期（Aalenian-Bajocian）］。

　　沙尔湖植物群真蕨类、银杏类、茨康类及落叶松柏类丰富，总体上显示了中国北方锥叶蕨 – 拟刺葵植物群（Coniopteris-Phoenicopsis flora）的面貌。但该植物群含有大量苏铁类（包括本内苏铁类），孢粉中也发现了一定数量的罗汉松粉（Podocarpidites）、里白孢（Gleicheniidites）和桫椤孢（Cyathidites）等偏热带—亚热带分子，显示了一定的中国南方早—中侏罗世植物群的特征（周志炎，1995）。因此，沙尔湖植物群也呈现了中侏罗世北方和南方植物群混生的特征。这可能与沙尔湖地区在地理位置上邻近早—中侏罗世西伯利亚植物区与欧洲—中国植物区的交界地带有关（Vakhrameev，1988）。因此，沙尔湖植物群的发现对研究早—中侏罗世中国古植物地理分区具有重要意义。

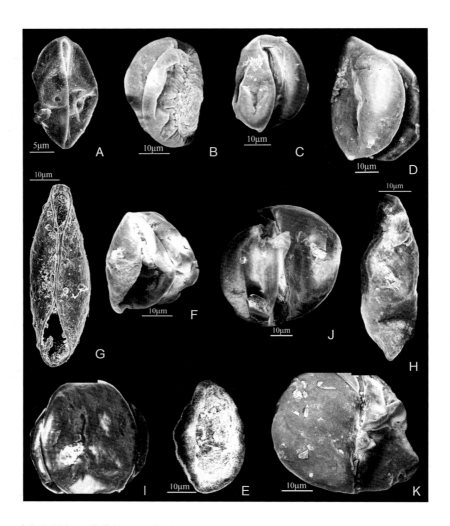

**图 48 沙尔湖植物群的花粉化石（2）**

Fig. 48 Pollen of Sherhu flora (2)

A. 具唇苏铁粉；B. 皱体双囊粉（未定种）；C、D. 雪松粉（未定种）；E. 无口器粉（未定种）；
F. 苏铁粉（未定种）；G、H. 卡城苏铁粉；I. 单束松粉（未定种）；J、K. 原始松柏粉（未
定种）（据董曼，2012）

A. *Cycadopites adjectus*; B. *Rugubivesiculites* sp.; C, D. *Cedripites* sp.; E. *Inaperturopollenites* sp.;
F. *Cycadopites* sp.; G, H. *C. carpentieri*; I. *Abietineaepollenites* sp.; J, K. *Protoconiferus* sp.（after
Dong, 2012）

就成煤而言，沙尔湖植物群丰富的银杏类、茨康类、苏铁类及落叶松柏类等植物为这里形成巨厚的工业煤层提供了充足的物质来源。沙尔湖地区蕨类植物的繁盛，反映了这里当时可能处于温暖潮湿的沼泽环境，且水体较为充沛。现今苏铁类主要生存在热带、亚热带地区，苏铁类及本内苏铁类的繁盛反映温度较高，易于植物生长；银杏类主要生存在亚热带及暖温带山间、排水良好的山地或平原地带，也反映当时的气候具有季节性变化。银杏类、茨康类、苏铁类及松柏类等可能都是良好的造煤植物。总之，中侏罗世早期，沙尔湖地区总体上可能处于温暖潮湿并具有季节性变化的暖温带的河湖边缘的湿地、沼泽或坡地。从沙尔湖煤田具有巨厚的煤层分析，这里中侏罗世时期应曾是茂密的森林，充足的造煤植物为形成大量的煤炭提供了保障。因此，本书将沙尔湖地区的植物群称为"中侏罗世沙尔湖森林"。

## 3.9 新疆中生代"十大化石明星"

经过本科研队 20 多年的研究，结合中国古生物学家徐星、汪筱林等近年来在研究新疆中生代化石取得的最新成果，本书遴选了以下 10 个分类群化石作为自 2000 年以来新疆中生代化石新发现的"十大明星"（图 49），其产地和产出层位为：

1. 新疆巨龙 *Xinjiangotitan*（鄯善，$J_3$）

2. 鄯善侏罗纪恐龙足迹群（鄯善，$J_2$）

3. 左氏准噶尔兽 *Dsungarodon zuoi*（牙齿化石，硫磺沟，$J_3$）

4. 乌苏新疆龟 *Xinjiangchelys wusu*（鄯善，$J_3$）

5. 水龙兽 *Lystrosaurus*（吉木萨尔大龙口，$T_1$）

6. 五彩冠龙 *Guanlong Wucaii*（五彩湾，$J_3$，据徐星等，2006）

7. 天山哈密翼龙 *Hamipterus tianshnaensis*（哈密，$K_1$，据 Wang et al.，2014）

8. 郝家沟南漳叶 *Nanzhangphyllum haojiagouense*（郝家沟，$T_3$）

**图 49 新疆中生代"十大化石明星"**

Fig. 49  Top Ten Star Fossils in Mesozoic Xinjiang

A. 新疆巨龙（鄯善，J₃）；B. 鄯善侏罗纪恐龙足迹群（鄯善，J₂）；C. 左氏准噶尔兽（硫磺沟，J₃）；D. 乌苏新疆龟（鄯善，J₃）；E. 水龙兽（大龙口，T₁）；F. 天山哈密翼龙（哈密，K₁，据 Wang et al.，2014）；G. 五彩冠龙（五彩湾，J₃，据 Xu et al.，2006）；H. 郝家沟南漳叶（郝家沟，T₃）；I. 奇裂（？）丁菲羊齿（郝家沟，T₃)；J. 沙尔湖似木贼（鄯善沙尔湖，J₂）。（H、I 据 Sun et al.，2001b，2010；J 据董曼，2012）

A. *Xinjiangotitan*, shanshan, J₃; B. Dinosaur track group, shanshan, J₂; C. *Dsungarodon zuoi*, Liuhuanggou, J₃; D. *Xinjiangchelys wusu*, shanshan, J₃; E. *Lystrosaurus*, Dalongkong, T₁; F. *Hamipterus tianshnaensis*, Hami, K₁, after Wang et al., 2014; G. *Guanlong wucaii*, Wucaiwan, J₃, after Xu et al., 2006; H. *Nanzhangphyllum haojiagouense*, Haojiagou, T₃; I. *Thinfeldia? mirisecta*, Haojiagou,T₃; J. *Equisetites shaerhuensis*, shanshan Shaerhu, J₂. （H, I after Sun et al., 2001b, 2010; J after Dong, 2012）

9. 奇裂（？）丁菲羊齿 *Thinfeldia? mirisecta*（郝家沟，$T_3$）

10. 沙尔湖似木贼 *Equisetites shaerhuensis* Dong（沙尔湖，$J_2$）

上述"十大明星"中，除哈密翼龙（汪筱林等发现）和五彩冠龙（徐星等发现）外，都是本科研队近年来在新疆中生代研究中的新发现。形体庞大的新疆巨龙、由恐龙足迹代表的食肉性兽脚类恐龙、早期哺乳动物准噶尔兽、两栖类乌苏新疆龟，以及形态特征独特的 2 个种子蕨分类群及 1 个有节类分类群等，共同展现了中生代时期新疆地区生物的丰富多彩，也展现了它们特有的发展及演化历程。

第 4 章

# 新疆化石情

## 4.1 白杨河化石之恋

托里白杨河谷是新疆西北准噶尔广袤戈壁沙漠中的一片绿洲。从白杨河以西的铁厂沟到白杨河以东的和什托洛盖，是约 200 千米的茫茫戈壁滩上唯一能找到的绿地，白杨林的生长也充分展现了生命的顽强。白杨河的神奇之处还在于，这里有丰富而又奇妙的化石，它们记载了 1 亿多年前生命演化在新疆准噶尔盆地的历史，也为揭开地质学家奥勃鲁契夫的"狄阿姆之谜"敞开了迷宫。20 多年来，白杨河畔留下中德科研队队员们艰辛的足迹和汗水，也留下他们许多难忘的记忆。每每回味，都使人感到无穷的眷恋。

### 4.1.1 七进白杨河

托里县的白杨河位于额敏县铁厂沟以东约 50 千米，自高耸的莎勒克腾山发源，先是湍急地向南奔流而下，到戈壁滩上的白杨河谷（即"阿克扎尔河谷"）时水流已变得平缓。夏季雨水少，可轻松地蹚水过河；但春季山雪融化或秋天雨季到来时，这里的水深会没过膝盖，要蹚过湍急的河水就有些艰难。主要产化石的地质剖面在河东，科研队驻地铁厂沟在河西，因此科研队队员每天清晨都要蹚水过河，从西岸去往东岸。河水冰冷，河底还有圆滚的砾石阻挡脚步，有时拄着木棍过河都不容易。

为了在这里采集更多的化石，进一步揭开当年奥勃鲁契夫的"秘密"，以孙革为首的科研队的中国队员们自1989年开始，曾7次来到这里。每次首先面临的困难都是过河：清晨从西向东过河时要带上全副野外装备，包括钢钎、铁锤、测绳和装满包装纸的纸箱等；晚上收工时，纸箱里装回的是一箱箱沉重的化石标本和岩石标本；有时，湍急的河水已没过膝盖，脚下随时可能被河底的砾石滑倒；遇到春秋两季，冰冷的河水不仅灌进靴子，有时还将整条裤子湿透。2015年10月，科研队第6次来白杨河时，由于采集的化石多，每只箱子的重量几乎都超过30千克。队伍中的梁飞博士最年轻，争着将最重的化石标本箱扛在肩上。他跌跌撞撞地将标本箱运过河后，还要返回帮助老师和同事们。"众人同行，小弟受苦"是中国的一句古语，意思是说，最年轻的人要多受累。梁飞正是实践这句名言，将多吃苦、热情助人作为工作中的乐事。

2018年5月初，科研队利用"五一"劳动节假期，第7次来到白杨河。由于山上的雪已部分融化，白杨河水已齐腰深，无法蹚水过河。为此，队员们决定改道：驱车前往距白杨河以东约200千米的和什托洛盖镇"安营扎寨"，每天由那里再折返，跨过茫茫戈壁滩再到达白杨河的东岸。这样一来，每天光路上往返就要花费6~7个小时，而且要克服戈壁滩上去白杨河根本没有公路等困难。为保证工作时间，大家早起晚归，每天收工返回驻地时都已夜幕降临。但是，所有这些困难都没有难倒年轻的科研队员们，大家主要想的是：珍惜宝贵时间，采集更多化石，力争又有新的发现（图50）。

### 4.1.2 "牙疼路"

白杨河剖面及化石点在铁厂沟以东约50千米的戈壁滩上，从铁厂沟向东北，先有一段大约30千米的柏油路，之后，当公路折向北去的时候，汽车需离开公路，驶入戈壁滩的碎石路一直向东。碎石路几乎无人光顾，道路的凸凹不平使汽车颠簸不断，戈壁滩上的碎石飞溅起来经常拍打着车门。

　　2015 年 10 月，当科研队来白杨河野外工作时，正赶上领队孙革教授牙病重犯，有时疼痛难忍。驻地铁厂沟是一个小镇，在这里要找牙医，得到 100 千米以外的克拉玛依市的医院。为了节省野外宝贵的工作时间，孙革教授只好忍痛坚持工作。但特别令人烦心的是，每当汽车驶离公路，驶上这段大约 10 千米的戈壁滩碎石路时，路上的颠簸顿时让孙革教授的牙疼得更厉害。有

**图 50　七进白杨河**
Fig. 50　Entering the Baiyang River valley for seven times

A. 托里白杨河（2015）；B~H. 在白杨河剖面野外工作，其中 C 为苗雨雁博士（2005），G 为孙革教授给年轻同事现场指导（2015），B、H 为涉过白杨河；I. 当地哈萨克牧民
A. Baiyanghe geological section; B–H. Field working in the Baiyanghe section: C showing Dr. Miao Y. Y. （2005），G showing Prof. Sun G. guiding the young colleagues, B, H showing crossing the Baiyang River; I. Local Kazakh herdsman

时车不得不停下来，等他吃点药，缓解一下疼痛再重新上路。当汽车离开这段颠簸的碎石路，驶入平缓的公路时，他的牙疼常常立刻就得到缓解。就这样，孙革教授一直忍着牙疼，带病坚持野外工作。由此，每当走到这段路时，大家都开始犯愁；离开这段路时，大家便顿时松一口气。孙革教授安慰大家说："找化石比牙疼重要；看到化石，牙疼就能减轻。"

从此，大家风趣地称白杨河西这段大约 10 千米的戈壁滩碎石路为"牙疼路"，并希望有机会时，能在这段路的起点设一个路牌，上书"驶过牙疼路，前方是通途"。

### 4.1.3 博士生的训练

科研队的苗雨雁研究员当年曾是孙革教授的博士生，本科就读中国地质大学（武汉），硕士和博士分别就读于中科院南京地质古生物研究所和吉林大学。在孙革教授的指导下，苗雨雁的硕士论文和博士论文的课题都是白杨河植物群研究。

2001 年至 2005 年，苗雨雁曾多次来白杨河开展野外工作，为硕士论文和博士论文收集化石资料并参加中德工作站的工作，同行的还有陈跃军教授、李春田、杜青工程师和程显胜高工等。4~5 月的白杨河，由于地处戈壁滩，早晚天气仍然很冷。早晨过白杨河时，上身穿着冬天的羽绒服，下身却是打着赤脚，卷起裤腿，在冰冷的河水中蹚过；到中午，头顶已是烈日炎炎，露头上的岩石已经发烫，有时风沙中还常常卷着小石子，一两天后，大家的脸庞都晒得黝黑，手上也磨出了大水泡。令人欣慰的是，科研工作取得丰厚收获，苗雨雁还从哈萨克牧民那儿学会了骑马。白杨河野外工作结束后，她顺利地完成了室内研究和论文写作，2006 年获博士学位。多年来，有关白杨河植物群，她已发表过多篇论文。每当回忆起在白杨河的野外工作，她都感慨万分。成为研究生之前，她虽也有多年的地质工作经验，但相比之下，准噶尔戈壁滩的工作格外难忘：在干热和风沙雕琢出的岩层间挥洒汗水，既是被大自然滋养的幸福享受，更是锤炼意志、增长学识的绝好机会。多年来，苗雨雁与白杨河的化石结下了深厚的情缘（图 50）。

## 4.2　三下鄯善戈壁滩

　　中德科研队合作研究工作的高潮出现在 2007~2013 年"三下鄯善戈壁滩"期间，队员们在吐哈盆地的鄯善七克台地区找寻恐龙及龟等化石。鄯善戈壁滩记录了中德科研队队员们不辞辛劳、执着找寻化石的可贵精神，也发生了许多动人的故事。

### 4.2.1　首进鄯善戈壁滩

　　从 2007 年起，中德科研队的重点工作转向在吐哈盆地寻找恐龙化石。这项工作的起因是 2006 年 11 月在德国法兰克福召开的中德联合实验室第二次会议。会议期间，时任新疆第一区调大队大队长左学义根据他多年在吐哈地区工作的经验，提议到吐鲁番地区寻找恐龙化石，认为可能会有新发现。2007 年 9 月初，恰好科研队在乌鲁木齐举办中德合作中国北方地质与环境演化国际学术研讨会，会后有野外考察，为开展此项工作提供了机会。于是，科研队成立了赴吐哈盆地工作的小分队，由德国青年专家维恩斯牵头，请科研队的中国恐龙学家董枝明教授指导，参加的队员大都是德国的年轻研究生，当然还有中方新疆第一区调大队工程师杜青陪同和协助。吐哈盆地是大家长久向往的地方，这次算是如愿以偿。小分队到达的第一站是吐鲁番以东约 90 千米的鄯善，这里曾是 20 世纪 70~90 年代董教授发现"中日蝴蝶龙"及"鄯善龙"等恐龙化石的地方（图 51）。

　　董枝明教授是国际著名恐龙学家，早年毕业于复旦大学，1963 年起作为中科院古脊椎动物与古人类研究所的研究生，师从中国古动物学家杨钟健院士，开始了他从事恐龙研究的学术生涯。半个多世纪以来，董教授足迹遍布中国大江南北特别是内蒙古及新疆等边远地区，先后发现了众多恐龙化石，被誉为"中国的恐龙王"。20 世纪 70~80 年代，他在吐鲁番地区发现白垩纪"鄯

**图 51 首进鄯善戈壁滩**

Fig. 51 Entering the Gobi in Shanshan for the first times

A. 中德新疆学术研讨会（乌鲁木齐，2007）；B~D. 鄯善戈壁滩，其中 B 示科研队进入鄯善戈壁滩（2007），C、D 为鄯善七克台侏罗纪地层；E. 董枝明教授（左）在鄯善野外指导维恩斯博士

A. The 3$^{rd}$ Sino–German symposium on Xinjiang（Urumqi, 2007）；B–D. Gobi of Shanshan in Turpan region: B showing the Sino–German research team entering the Gobi of Shanshan（2007）, C and D showing the Jurassic strata in Qiketai of Shanshan; E. Prof. Dong Z. M. (left) guiding Dr. Wings O. in Shanshan field work

善龙"和"嘉峪龙"（1977）及侏罗纪"中日蝴蝶龙"（1992）等重要恐龙化石。本次率领小分队来鄯善，也是他"故地重游"，希望在这里有新发现。除董教授外，小分队中方的杜青工程师是多年从事新疆地质工作的专家，他早年毕业于西北大学地质系，多年的野外工作使他积累了丰富的经验。他身体彪悍健壮，野外工作中还能保护小分队队员们的安全。德方青年学者维恩斯博士是恐龙学专家，毕业于德国图宾根大学地质古生物学院，当时在柏林自然史博物馆工作，年富力强、勇于开拓。维恩斯的同伴包括来自波恩大学的谢里洪（Schellihorn R.）、科纳（Koerner K.）等研究生，他们虽初次来新疆，但对找寻恐龙化石充满了憧憬；他们在德国已有多次野外化石采掘的经验。

9 月的鄯善戈壁滩骄阳似火，戈壁滩的南面便是中国八大沙漠之一的库木塔格沙漠，绵延约 400 千米。炎热的戈壁滩地面犹如被火炉烘烤，气温有时可达近 50 ℃。董教授带小分队来到鄯善以东约 20 千米的七克台，这里绵延数十公里的戈壁滩上，紫红色及灰绿色的岩石呈条带状铺开，大量黝黑色的砾石杂乱无章地被覆在戈壁滩上，一片荒漠无垠的枯燥。小分队沿南北向的沟壑，一条沟一条沟地穿梭式找寻 20 多年前董教授发现的恐龙化石点，犹如大海捞针。但不管怎样，董教授已为德国的年轻人指明了方向。

就这样，在董教授从鄯善离开后，维恩斯和其他几位德国年轻助手在杜青工程师的陪同下又在戈壁滩上顽强地工作了几天。"功夫不负有心人"，到野外工作的第 4 天，他们终于在鄯善七克台南部的一个狭窄的沟壑里有了新发现——首次发现侏罗纪恐龙足迹群。"一下吐鲁番鄯善戈壁滩"首战告捷。

## 4.2.2 意外的午餐

对于董教授离开鄯善，德国年轻队员们有些依依不舍，但离开了中国老专家帮扶的大手，几位德国年轻人又多了一份自信。在董教授离开鄯善的第二天，一大早，维恩斯就带领几个年轻同伴，在杜青工程师的陪同下，又一次踏上鄯善七台河以南的戈壁滩。他们从南到北、从西向东，先后穿过几个由杂色和红色岩石构成的南北向沟壑。数十次爬上爬下，他们已经有点筋疲力尽了。眼看中午到了，他们决定在一处略开阔的沟壑边休息并吃午饭（图52）。

几个年轻人上了沟东坡，坐在土坡上开始吃饭，一边嘴里嚼着香肠和面包，一边眺望或审视着戈壁滩中沟壑的景色。中午的太阳十分强烈，但阳光照耀下，沟壑的轮廓也十分鲜明，特别是对面坡崖上半黄半绿的岩石露头在阳光下色彩斑斓，很像一幅岩石壁画。突然间，一个年轻人放下饭盒，急匆匆向对面坡上的土状石崖跑去，原来，他无意间望见石崖上的半黄半绿色的岩石露头上有凸起物，有些新奇。大家诧异地望着他怪异的举动。忽然，年轻人在石崖上惊呼："呵！快来呀！这里有新发现！"

大家立刻放下手中的饭盒，奋力向对面的斜坡跑去。仔细一看，果然是一批密集分布的、三叉状的凸起暴露在黄绿色的岩石上！"是恐龙足迹！"维恩斯肯定地说，并招呼大家行动起来。大家十分振奋，一起顺着印有足迹的岩层方向，向北追索过去。果然，在一面更大的、长约30米的岩石露头上又发现更多的恐龙足迹化石，加上北面又发现的足迹化石点，三处足迹化石加起来共有130多枚！就这样，一个奇特的、新疆侏罗纪最大的恐龙足迹化石群在这里被偶然发现，发现时间就在这顿戈壁滩上的午餐时刻！

很快，维恩斯将此项重要的恐龙足迹化石发现的消息报告给科研队领队孙革教授。孙革闻讯后，赶紧带吴文昊博士等助手专程赶到现场。后来，董枝明教授等专家也赶来了。专家们均确认，这是中国首次侏罗纪恐龙足迹的新发现。紧接着，后续研究紧锣密鼓地开展起来，包括对恐龙足迹化石群的描述、分类及古生态等研究。成果论文由维恩斯牵头、董枝明教授、吴文昊副教授等参加，很快于中国的《世界地质》（Global Geology）杂志发表。

2008年4月8日，由中德科研队牵头，鄯善县政府在鄯善举行了新闻发布会，正式宣布了迄今世界最大的侏罗纪恐龙足迹群在鄯善的首次发现。科研队领队孙革和德方代表马丁教授分别在会上介绍情况，北京自然博物馆的恐龙足迹研究专家李建军研究员等也应邀出席，并对此次足迹化石的发现予以高度评价。新华社、新疆当地的各大媒体的记者齐聚会场，迅速报道了此

**图 52　戈壁滩上意外的午餐**

Fig. 52　Unexpected lunch in Gobi

A. 鄯善七克台戈壁滩上的足迹化石点及午餐地点；B. 对面足迹化石；C~E. 三个足迹化石点及野外工作；F. 维恩斯博士

A. Dinosaur tracks fossil site and lunch place in Gobi of Shanshan; B. Dinosaur tracks found in beginning; C–E. Three main outcrops of dinosaur tracks and the working; F. Dr. Wings O.

**图 53 鄯善恐龙足迹群研究成果新闻发布会**

Fig. 53 Press on discovery of dinosaur tracks in Shanshani

A. 董枝明教授介绍鄯善侏罗纪恐龙足迹化石；B. 德国专家维恩斯（左 1）及中国专家李建军（左 2）介绍恐龙足迹化石；C. 孙革教授在新闻发布会上做介绍；D、F. 会议代表在恐龙足迹现场；以上均为 2008 年；E. 莫斯布鲁格院士（中）考察恐龙足迹群产地（2009）；G、H. 恐龙足迹化石

A. Prof. Dong Z. M. introducing the dinosaur tracks; B. Experts introducing the dinosaur tracks including Dr. Wings（left 1）and Prof. Li J. J.（left 2）; C. Prof. Sun G. introducing the discovery to the Press in Shanshan; D, F. Participants at the dinosaur track fossil site; (all above in 2008）; E. Prof. Mosbrugger（mid）looking at the dinosaur track site （2009）; G, H. Dinosaur tracks

**图 55　新疆鄯善恐龙之声**

Fig. 55　The Sound of Dinosaurs in Shanshan, Xinjiang

A~F. 新疆巨龙化石发掘现场，B 为新疆巨龙部分骨骼化石；G. 新疆巨龙发现新闻发布会（2012）；H. 董枝明教授在发掘现场介绍新疆巨龙化石；I. 德国波恩大学专家参观化石发掘现场（2013）

A–F. Excavation of *Xinjiangotitan*, B showing parts of the bones of *Xinjiangotitan* in situ; G. Press conference on the discovery of *Xinjiangotitan* in Shanshan（2012）; H. Prof. Dong Z. M. introducing the dinosaur ; I. Experts of Univ. Bonn, Germany visiting the fossil site（2013）

*shanshanensis*）；新属名赠予新疆，种名赠予鄯善，作为中德科研队送给新疆鄯善最好的礼物。2013 年 9 月，"中国最大的侏罗纪恐龙在新疆鄯善首次发现"的研究成果在《世界地质》正式发表。后续发掘工作在恐龙学家李大庆教授率领下又取得重要进展：除发现恐龙的部分头骨外，又发现 15 节颈椎（共 18 节全部完好保存）及 39 节尾椎等，进一步提高了研究程度，成果于 2018 年在国际学术期刊《历史生物学》（*Historical Biology*）发表（Zhang et al., 2018）。新疆巨龙这一新发现首次展示了中国最大的侏罗纪恐龙，刷新了中国侏罗纪大型恐龙的纪录，对深入研究中国恐龙动物群的组成与演化，以及新疆侏罗纪古地理、古气候等具有十分重要的意义，也为鄯善化石保护、建立国家地质公园及地质博物馆等做了难能可贵的科学铺垫。至此，"三下鄯善戈壁滩"终于圆了大家的"鄯善恐龙梦"，科研队全体队员们似乎已听到 1.6 亿年前生活在吐鲁番鄯善的新疆巨龙惊天动地的吼声（图 55）。

### 4.3.2　鄯善的节日

为祝贺迄今中国最大的侏罗纪恐龙——新疆巨龙在鄯善的首次发现，也为推动鄯善及新疆化石保护工作的进一步开展，2013 年 10 月 10 日，鄯善县政府在鄯善隆重举办了新疆巨龙命名仪式暨化石保护现场会，也是一个庆功大会。国家古生物化石专家委员会、新疆维吾尔自治区国土资源厅及吐鲁番地区政府的领导、中德科研队的部分成员、德国专家马丁教授和里特（Litt T.）教授和他们带领的波恩大学地质古生物专家团队，以及鄯善当地群众等，共 100 余人出席了大会。会上，鄯善县群众向本次在鄯善发现恐龙化石的有功专家们献花，时任县委书记代表县委及县政府发表了热情洋溢的讲话，祝贺新疆巨龙的成功发现并感谢中德科研队的辛勤工作。会场上，彩旗飘扬，五颜六色的气球在空中飞舞，鄯善人民在欢迎一个喜庆、振奋人心的节日（图 56）。

为了进一步宣传古生物化石在科学研究和科学普及中的重要作用，更好地将科研成果转化为鄯善恐龙化石保护及科学普及服务，在中德科研队中方

**图 56 鄯善的节日及建设鄯善侏罗纪博物馆**

Fig. 56 Festival for Shanshan and building the Jurassic Museum of Shanshan

A. 祝贺鄯善新疆巨龙命名大会（2013）；B. 建设中的鄯善侏罗纪博物馆；C~E. 部分主要展品，其中 C 为新疆巨龙，D 和 E 分别为鄯善珍贵矿石钠硝石及鄯善玉；F. 沈阳师范大学古生物学院师生来恐龙化石现场野外实习（2016）；G. 德国青年专家为博物馆设计的广告场景

A. Celebration for naming *Xinjiangotitan* (2013)；C. The Jurassic Museum of Shanshan in construction；C–E. Some main exhibits of the Museum: C showing the dinosaur *Xinjiangotitan*; D, E. Precious minerals in Shanshan, D showing Soda niter, E showing Shanshan Jade; F. Students and teachers of College of Paleontology, SYNU visiting the dinosaur site for geological excursion (2016)；G. Design of the Shanshan Museum presented by German young scientist

领队孙革教授提议下，2016年7月16日，由国家古生物化石专家委员会办公室牵头，在鄯善举行了建设鄯善地质古生物博物馆专家论证会。与会专家一致赞同，尽快在鄯善建立地质博物馆，并提议博物馆命名为"鄯善侏罗纪博物馆"，以彰显鄯善在侏罗纪恐龙等化石蕴藏丰富、在化石研究中取得了国际一流成果的特色。孙革教授亲自为鄯善这一新建的博物馆撰写了展陈大纲，得到与会专家评委的高度评价。

春华秋实，中德科研队科学家们的"化石情"终于结下了丰硕的果实。每当回忆起在新疆与化石在一起的日子，队员们的心中不仅深深留恋，也感到无比自豪。

第 5 章

# 中德友谊之花

## 5.1 "第二次生命"

中德合作科研队在新疆最难忘的故事莫过于科研队的核心成员、德国阿什拉夫教授的"第二次生命"了。"阿教授"（中国新疆同事们最喜欢这样称呼他）曾在新疆野外工作中意外受伤，被奋力抢救后转危为安的故事已传为佳话。阿教授常说："中国新疆给了我第二次生命！"

阿什拉夫是德国孢粉学家，祖籍阿富汗，是已故阿富汗国王查希尔（Zahir Shah M.）的侄子。1966 年毕业于喀布尔大学地质系，1967 年举家迁往德国，1972 年于德国波恩大学获学士学位，1977 年获博士学位，师从德国著名古生物学家沃斯特（Wurster P.）和斯崴瑟教授。1978 年起，先后于波恩大学古生物研究所和图宾根大学地质古生物学院从事科研工作，1989~1991 年受聘菲律宾大学迪利曼分校客座教授，2001 年被聘为中国吉林大学名誉教授，2002 年起任吉林大学古生物学与地层学研究中心（RCPS）学术委员会委员，2004 年起兼任中德合作新疆地质工作站副主任。多年来，他在德国泥盆纪及第三纪孢粉和地层研究、中亚地区北伊朗和阿富汗地区早中生代孢粉及地层研究、中国粤北南雄第三纪孢粉及 K/T 界线研究、黑龙江嘉荫晚白垩世生物群及 K-Pg

界线研究，以及新疆准噶尔盆地中生代孢粉及地层研究等工作中，均作出重要贡献。他多次访问中国，在中德科学合作中一直发挥重要的作用，并对中国和中国人民怀有深厚的感情，他的性格和人品兼具了东、西方特有的优点（图57）。

1997年7月，阿什拉夫以主要成员身份参加了由孙革教授和莫斯布鲁格教授主持的中德马普"新疆中生代准噶尔盆地生物地层合作研究"项目。7月9日下午，科研队在郝家沟野外工作结束后准备返回乌鲁木齐的途中，路过一个已废弃的旧煤矿坑口时，他要汽车停下来，给陪同的德国博士研究生乌尔(Uhr D.)做现场煤层地质指导。万万没料到的是，这个旧坑口已关闭多年，木制的坑口支架早已松垮，支架上面巨大而松散的石块随时都有塌落的危险。阿什拉夫当时只顾给研究生讲解，没注意到这里的安全隐患，当他用地质锤触指顶上的岩石介绍其地质特征时，上面的巨大石块突然塌落，几块落石将他的身体夹住并砸伤。阿什拉夫顿时感到臀部剧烈疼痛，很显然，他的胯部已经严重骨折。见此情景，研究生乌尔立刻呼叫救人，科研队在车上等待的同事们都跑了过来紧急解救。但巨大的岩石实在太大太重，几个人都难以搬动。最后，还是附近的煤矿工人们闻讯赶来协助，众人一起奋力用木棍撬挪，才把阿什拉夫从大石块下救出。为了马上将阿什拉夫送乌鲁木齐医院抢救，当地煤矿工人们现做了一副临时担架，让阿什拉夫能平卧在汽车上，由于他太重，最后6个人合力才将他抬到车上，紧急送往乌鲁木齐医院。郝家沟至乌鲁木齐在矿区这段路大约20千米，但是戈壁滩上都是凸凹不平的土路，为了减少颠簸对阿教授伤痛的影响，这段路程汽车竟然行驶了将近4个小时！待到乌鲁木齐时，已近夜深，汽车直奔乌鲁木齐最好的医院——新疆军区陆军总院。

当天，领队孙革教授因病未能出野外，晚上一直焦急地坐在招待所的门口的石阶上等待大家收工。得知阿教授受伤的消息，孙革赶紧奔赴陆军总院，先是去急诊室放射科看片子，确认了阿什拉夫是胯骨骨折，之后立即赶到病房，查问病情。孙革与阿什拉夫是多年亲密挚友，两人见面时同时流下了热泪。当晚，孙革当即给南京打电话，请人前来增援。第二天，南古所年轻的

**图 57　德国孢粉学家阿什拉夫和他的"第二次生命"**

Fig. 57　German palynologist Ashraf with his "Second Life"

A. 阿什拉夫教授；B~D. 野外工作地点及塌落岩石；E. 新疆陆军总院护士们照料阿什拉夫
（1997）；F. 德国保险公司（ADAC）专机接阿什拉夫回德国接续治疗，图为在乌鲁木齐
机场抬阿什拉夫上飞机回德国进一步治疗；G. 阿什拉夫在飞机上；H. 阿什拉夫在中国讲学；
I. 阿什拉夫在德国柏林接待中国同行（2013）；J. 孙革教授在新疆乌鲁木齐送阿什拉夫书
法赠品

A. Prof. Ashraf; B–D. The field working place and the accidentally falling down rocks; E. Ashraf in
the Army Hospital of Urumqi looked after by Chinese nurses（1997）; F. Carrying up Ashraf into
the ADAC plane for return to Germany for further treatment; G. On the plane （1997）; H. Ashraf
teaching students in China; I. Ashraf receiving Chinese colleagues in Berlin, Germany
（2013）; J. Sun G. presenting calligraphy to Ashraf

王怿博士乘飞机从南京紧急赶来，与科研队博士生王鑫甫轮流看护阿什拉夫。医院检查结果表明：阿什拉夫的右侧胯骨骨折错位约 2 厘米，属于严重骨折。为了使阿什拉夫得到及时有效的医治，陆军总院调出单人病房给他，医生和护士们也都热心照顾（图 57，E）。由于当时先进的通信设备在中国还不普及，阿什拉夫与家人通信联系不便，新疆第一区调大队左学义队长便找来手机让阿什拉夫打国际长途电话，与远在瑞士的母亲联系，看到他同母亲讲话时泪水不停地流下，在场的同事们都感到无比的心疼。在阿什拉夫住院的十天里，新疆第一区调大队的领导和同事们都前来看望，新疆科委招待所的彭莲花所长还特地给他送来鲜花，这里的病房犹如一个温暖的大家庭。中国同事和医院对阿什拉夫的关心和友谊令他十分感动。

为保证阿什拉夫的骨伤能尽快痊愈，在德方领队莫斯布鲁格教授努力下，德国保险公司（ADAC）派专机和 4 名工作人员（2 名医生和 2 名驾驶员）从德国飞来乌鲁木齐，接他回国治疗。为办理保险公司的手续，孙革和莫斯布鲁格分别在中德两国为德国保险公司提供资料，孙革等还带人去乌鲁木齐机场送行。在机场，当阿什拉夫的担架置放在停机坪旁，送行人与他亲切告别时，大家都热泪盈眶。眼望德国飞机渐渐远离乌鲁木齐，中国同事们都衷心地遥祝他一路平安。最终，阿什拉夫平安返回德国并在图宾根大学医院精心治疗下痊愈，所有中国同事都为他的转危为安感到无比的庆幸。

由于此次野外安全事故发生在 7 月 9 日，而阿什拉夫的生日是 7 月 2 日，从此，阿什拉夫就将每年的 7 月 9 日作为他"第二个生日"。后来，每当他再来到中国时，逢人便讲，是中国新疆给了他"第二次生命"。2000 年 12 月，当阿什拉夫身体康复重返新疆时，孙革教授在乌鲁木齐的欢迎会上向他赠送了自己亲笔书写的"新疆夏日长，兄弟手足情"条幅，代表科研队中国全体同事向他表达了对他有幸脱险的衷心祝福（图 57，J）。

# 5.2 中德"大家庭"

"来到乌鲁木齐，我们又回到了第二个家！"这是科研队德方队员们每次重返新疆时总要说的一句动情的话。的确，在新疆地矿系统领导和同事们的热诚努力下，中德合作新疆地质工作站成了中德两国科学家温暖的大家庭。总部设在新疆第一区调大队的这个"大家庭"，多年来为中德两国地学合作做了大量工作，使中德两国科学家不仅在学术上紧密合作，也建立起深厚友情（图 58，图 59）。

为了安排好中德工作站的工作，东道主新疆第一区调大队专门腾出一座办公楼作为工作站的办公地点，并派古生物专业的李洁工程师等专门负责接待工作（图 58，C）。为方便野外工作，区调大队准备了多辆越野车，都是当时国内最好、最安全的车型。中德科研队工作的第一站是距乌鲁木齐市区约 50 千米的郝家沟，每天出城向西后，还要一个多小时才能到达。进入郝家沟之前大约 10 千米都是年久失修的土路，每次遇到陡坡汽车难行，大家都下来推车或改为步行。工作起点的郝家沟地处偏僻，这里仅住有一户叫哈德力的哈萨克牧民，家里有兄弟二人。他们看到中德两国的科学家不辞辛苦地在山野工作，很受感动，见到科学家们采集的石头样品太多、太重，哈德力兄弟总是牵着骆驼、马或驴前来协助。时间长了，他们的骆驼也与科研队的汽车成了"好朋友"。

中德合作工作站在新疆十年间（2000~2009），德国专家（包括带学生）多次前来新疆工作或野外实习。其间，中方专家（包括新疆第一区调大队的专家）也应邀多次访问德国，先后赴图宾根大学、波恩大学和森肯堡博物馆等。为了培养人才，在莫斯布鲁格教授的厚意安排下，新疆第一区调大队年轻的办公室主任郭红也应邀到德国图宾根大学短期进修，在德国学习了一个月，既扩大了国际视野，也增长了才干。后来，她被抽调到新疆科委 305 办公室担任更重要的工作。中德两国专家在新疆的"大家庭"中，也同新疆当地少数民族群众建立了深厚的友谊。

**图 58 中德"大家庭"**
Fig. 58 Sino-German "Big Family"

A. 中德科研队在天山天池（2001）；B. 中德学术研讨会在法兰克福举行（2006）；C. 中德工作站"大家庭"的"夫妻管家"杜青与李洁高工；D. 莫斯布鲁格与孙革等在鄯善维吾尔族群众家中做客（2009）；E. 孙跃武教授为德国同行烤羊肉串做午餐（硫磺沟，2005）；F. 在莫斯布鲁格家中做客，右 3 为莫斯布鲁格夫人伽比（2012）；G. 阿什拉夫与中国新疆维吾尔族老人亲切交谈（吐鲁番，2007）；H. 新疆第一区调大队接待阿什拉夫教授（乌鲁木齐，2005）

A. Sino-German research team in Tianchi Lake of Tianshan Mt.（2001）；B. Sino-German symposium in Frankfurt, Germany（2006）；C. The couple stewards of the "big family"（Li J. and husband Du Q.）；D. Mosbrugger and Sun G. visiting the Uyghur Chinese people's home in Shanshan of Turpan （2009）；E. Prof. Sun Y. W. making mutton shish kebab for German colleagues' lunch in Liuhuanggou（2005）；F. Chinese colleagues invited in Mosbrugger's home（right 3: Ms. Gabi Mosbrugger, in Frankfurt, 2012）；G. Ashraf talking with the Uyghur Chinese old man in Turpan（2007）；H. Prof. Ashraf invited by the RGSX No. 1, in Urumqi（2005）

**图 59　中德科研队在新疆**

Fig. 59　The days for the Sino-German Research Team in Xinjiang

A. 莫斯布鲁格教授代表中德合作队向新疆第一区调大队献锦旗（1997）；B. 参观新疆第一区调大队新仪器（1997）；C. 阿什拉夫与孙跃武教授研究工作；D~F. 硫磺沟野外工作，其中 F 为哈萨克牧民哈德力兄弟协助科研队搬运岩石样品，E 示当地骆驼与科研队汽车成为好伙伴（2005）；G. 莫斯布鲁格教授与孙革教授在鄯善与维吾尔族化石保护人员（中）合影（2009）；H. 鄯善县领导（左3）与科研队员在恐龙化石点（2011）；I. 德国专家与当地少数民族群众合影（2013，大龙口）

A. Prof. Mosbrugger on behalf of the Sino-German team presenting flag to RGSX No. 1（1997）; B. Visiting the new facility of the RGSX No. 1（1997）; C. Prof. Ashraf discussing work with Prof. Sun Y. W.; D-F. Field work in Liuhuanggou: F showing the Kazak Chinese Khadli brothers helping in moving the rocks, E showing the camels being friends with the team's car; G. Prof. Mosbrugger and Prof. Sun G. with local protection personnel in fossil site in Shanshan Gobi（2009）; H. Local leader Ms. Zhou（left 3）with the research team members in the dinosaur site in Shanshan（2011）; I. The German experts with the local people in Dalongkou section （2013）

## 5.3 年轻人的成长

在新疆，人们最推崇的植物是胡杨，将其誉为"沙漠之神"。胡杨在恶劣的自然条件下能根深叶茂，充分展现了生命的顽强。中德科研队的年轻人在新疆的成长有如沙漠中的胡杨，他们在艰苦的野外工作中自觉锻炼成长，知识越来越丰富，能力越来越强，意志越来越坚定。

中德新疆中生代合作研究的开展，促进了一批优秀的年轻地质古生物工作者茁壮成长，如中国的苗雨雁、柳蓉、吴文昊、续颜、杨涛、董曼、边伟华、张建光、孟庆涛、周长付，以及日本在华博士生关谷透等。中德合作之初，他们大都是硕士生或博士生，如今他们均已获得博士学位，并成为各自研究领域的青年学术带头人。德国方面，青年专家麦什、马茨克、维恩斯、乔伊斯、韦丽娜及来自匈牙利的德国博士生拉比等，他们都曾在中德新疆合作研究的野外工作中得到锻炼；许多重要的古生物化石（如鄯善侏罗纪恐龙足迹群、新疆龟化石群等）也都是由他们在辛勤的野外工作中最早发现。他们先后在国际重要学术刊物，如《古脊椎动物学杂志》（JVP）、BMC、《自然科学》等共发表了20余篇高水平论文，充分展现了他们各自在专业上的成长。当然，他们论文的化石和地质材料都来自参加中德新疆合作的工作的实践（图60）。

中德新疆地质古生物的合作研究也为中国的年轻学者们赴德国深造创造了有利条件：苗雨雁、柳蓉、吴文昊、续颜、董曼、孟庆涛等都曾在德国短期进修，边伟华、张建光等在德国达姆施塔特大学获得博士学位。边、张的两位德国指导老师（辛德勒和胡侬教授）对来自中国的研究生们给予热忱的关怀和指导（图60，C，I）。博士生孟庆涛当年（2004）还充满稚气，她勇敢地站在乌鲁木齐的中德新疆国际会议上用英文演讲，受到与会专家的青睐，之后她得到科研队德国年轻孢粉学家布鲁荷（Bruch A.）等的热心指导并赴德国进修，业务上飞速成长。后来，她将油页岩的古环境研究与孢粉化石研究紧密结合，在古近纪油页岩研究工作方面取得突出成绩，现已成为吉林大学油页岩领域年轻的学术带头人之一。

**图 60 中德合作科研队的年轻人的成长**

Fig. 60 Growing of young scientists of Sino-German Research Team in Xinjiang

A、B. 苗雨雁、柳蓉博士在德国进修；C. 边伟华（右）与张建光在德国攻读博学位时与导师胡侬教授（中）在野外；D. 孟庆涛在中德会议上作报告（2004）；E. 德国研究生韦立娜在鄯善修复化石；F. 德国学生在硫磺沟野外实习；G. 德国研究生在硫磺沟采集龟化石；H. 中德年轻队员人在鄯善恐龙化石点；I. 中德联合培养博士生和他们的中德导师们；J. 吴文昊博士（右）与周长付博士在鄯善发掘恐龙；K. 德国青年专家（左起）乔伊斯、维恩斯和拉比博士在新疆

A, B. Maio Y., Liu R. in Germany for advanced studies; C. Bian W. H. and Zhang J. G. for Ph. D. study in Germany, with supervisor Dr. Hornung J.（mid）; D. Meng Q. giving talk in symposium in Xinjiang（2004）; E. German graduate Verena preparing fossil in Shanshan; F. German students in Liuhuanggou field work；G. German graduate collecting turtle fossil in Liuhuanggou; H. German and Chinese young scientists in the dinosaur site of Shanshan; I. Sino-German co-supervisors training Ph. D. students in field in China; J. Dr. Wu W. H.（right）and Dr. Zhou C. F. in Shanshan; K. German young experts（from left）Joyce W., Wings O. and Rabi M. in Xinjiang

# 5.4 中德双方授誉

以新疆中生代研究为主的中德合作的开展，带动了中德双方科研与教学工作，双方科学家也在对方国家获得许多荣誉。20多年来，莫斯布鲁格院士和阿什拉夫教授先后受聘中国吉林大学和沈阳师范大学名誉教授或客座教授；马丁教授受聘吉林大学和沈阳师范大学客座教授；吉林大学和沈阳师范大学分别设立了以阿什拉夫教授名字命名的"阿什拉夫实验室"；莫斯布鲁格、阿什拉夫和马丁等还分别被聘请担任中国学术刊物《世界地质》的副主编及编委等（图61，图62）。

中方科学家在德国授誉方面，2018年7月3日，德国森根堡自然博物馆在法兰克福广场（The Squaire）举行隆重集会，授予孙革教授德国森根堡自然科学协会通讯会员（Corresponding Member，SGN）荣誉称号；德国波恩大学地质古生物所向孙革教授颁发了荣誉证书。森根堡自然博物馆赫科纳博士、马丁及阿什拉夫教授等出席。莫斯布鲁格院士代表森根堡自然科学协会向孙革教授颁发了荣誉证书，并介绍了孙革教授在古生物学研究领域取得的成就及30余年来为中德两国地学合作作出的贡献（图61，G）。马丁教授代表德国波恩大学地质古生物所向孙革颁发了荣誉证书（图61，H）。德国森根堡自然科学协会成立于1817年，是享誉国际的著名学术团体之一；以往获SGN通讯会员和荣誉会员称号的包括歌德（von Goethe J. W.）、达尔文（Darwin C.）、居维叶（Cuvier G.）、黑格尔（Hegel G.）以及洪堡（Humboldt A.）等著名科学家，孙革是唯一获此殊荣的中国学者。这一荣誉的授予体现了德国科学界对中德合作的高度重视，也是中德友谊的一个象征。

此外，2010年10月，中德科学中心在北京举行成立十周年庆典，特邀孙革教授和莫斯布鲁格院士联合做大会唯一的工作报告，这也是中国国家自然科学基金委员会（NSFC）、德国科学基金委员会（DFG）对本科研队及中德地质古生物合作的高度重视和礼遇（图62，B）。

**图 61　中德双方授誉（1）**

Fig. 61　Honorary for Chinese and German two sides (1)

A. 莫斯布鲁格受聘中德地质古生物实验室主任（德方），中为时任吉林大学校长周其凤院士，左为中德科学中心主任韩建国教授（2005）；B、C. 莫斯布鲁格受聘沈阳师范大学名誉教授（2019）；D. 阿什拉夫受聘吉林大学客座教授（2004）；E. 阿什拉夫受聘沈阳师范大学名誉教授（2012）；F. 马丁受聘沈阳师范大学客座教授（2013）；G. 孙革受聘德国森肯堡自然科学协会通讯会员（法兰克福，2018）；H. 马丁教授代表波恩大学地质古生物所向孙革赠送证书；I. 森肯堡自然史博物馆代表等出席仪式（2018）

A. Mosbrugger V. appointed as Co-Director of Sino-German Lab of Paleontology & Geosci., Ex-President of Jilin Univ., Academician Zhou Q. F.（mid）and Co-Director of SGSPC, Prof. Han J. G.（left）in presence（2005）；B, C. Ceremony for Mosbrugger as Honorary Prof. of SYNU（2019）；D. Ceremony for Ashraf（left 2）as Guest Prof. of Jilin Univ.（2004）；E. Ceremony for Ashraf（left 2）as Honorary Prof. of SYNU（2012）；F. Ceremony for Martin as Guest Prof. of SYNU （2013）；G. Ceremony for Sun G. as Corresponding Member of SGN in Frankfurt, Germany （2018）；H. Prof. Martin presenting thankful paper of the IGMP of Univ. Bonn to Sun G.; I. Representatives of Senckenberg Museum（NH）joining in the ceremony（2018）

**图 62 中德双方授誉（2）**

Fig. 62 Honorary for Chinese and German two sides (2)

A. 莫斯布鲁格应邀担任"黑龙江嘉荫 K–Pg 界线国际学术研讨会"副主席（2019）；B. 孙革与莫斯布鲁格应邀出席"中德科学中心成立十周年纪念大会"（北京，2010）；C、D. 嘉荫地质公园设阿什拉夫塑像，阿什拉夫获"为嘉荫作出突出贡献的优秀科学家奖"（2017）；E. 阿什拉夫受聘沈阳师范大学名誉教授，时任校长赵大宇（左）颁发证书（2012）；F. 吉林大学古生物中心设立的"阿什拉夫实验室"；G、H. 辽宁古生物博物馆实验室展陈阿什拉夫孢粉照片（G）并建"阿什拉夫实验室"（H）；I. 莫斯布鲁格与孙革共同主持新疆学术研讨会（乌鲁木齐，2004）；J. 马丁受聘沈阳师范大学客座教授（2013）；K. 阿什拉夫（前排右1）应邀出席首届黑龙江 K–Pg 界线研讨会（长春，2002）

A. Mosbrugger invited as Vice–Chairman of the Int'l Symposium on the KPgB in Jiayin of Heilongjiang（2019）；B. Sun G. and Mosbrugger invited to presence "the 10th Anniversary of SGSPC" in Beijing（2010）；C, D. Ashraf statue built in Geopark of Jiayin, and he receiving the Award of Excellent Scientist for Jiayin（2017）；E. Ashraf awarded Honorary Prof. of SYNU（2012）；F. Ashraf Lab established in RCPS of JU; G, H. PMOL exhibiting Ashraf's palyno–fossil photos and establishing Ashraf Lab; I. Mosbrugger and Sun G. co–chairing the symposium in Urumqi, Xinjiang（2004）；J. Martin T. awarded the Guest Prof. of SYNU（2013）；K. Ashraf invited to attend the 1st Symposium on the KPgB in Jiayin of Heilongjiang（Changchun, 2002）

## 5.5 友谊之花香飘满园

多年来，中德两国已有 13 个科研机构参与新疆地学合作研究，中德友谊也不断加强并向外传播。双方新疆合作研究的不断扩大，吸引了美国、法国、日本、奥地利、匈牙利、卢森堡以及爱尔兰等多国科学家的参与。新疆中生代合作研究的成果已在国际上被广泛传播和引用，中德新疆友谊之花向全世界开放（图 63）。20 多年来，中德合作科研队得到美国科学院迪尔切院士（Dilcher D. L.，NAS）、日本中央大学古植物学家西田治文（Nishida H.）教授等热情支持与参与。

与此同时，中德新疆合作也为中德两国大学、自然史博物馆等的合作与交流架起了桥梁。2014 年，"中国带毛恐龙特展"在德国斯图加特自然史博物馆隆重举行，双方博物馆的领导都是本科研队的成员，展览在德国斯图加特持续了三个月，得到德国各界的广泛欢迎（图 63，D、G）。2017 年，德国波恩大学 26 名师生来沈阳师范大学和吉林大学联合开展野外地质实习，德国师生不仅在辽西和吉林等地参观了野外地质剖面，还在黑龙江嘉荫采集化石并出席了"纪念中国恐龙发现 115 周年大会及嘉荫化石保护论坛"，德国师生高度评价中德两国地学合作的成绩，并深受两国科学家深厚友谊的感染。

此外，两国科学家与新疆当地群众特别是居住在化石产地附近的少数民族群众也增进了友谊。2009 年 9 月，中德合作科研队的领队莫斯布鲁格和孙革在鄯善县领导的陪同下，应邀在鄯善当地维吾尔族家中做客，家中的女主人热情地拿出新疆著名特产——吐鲁番葡萄和哈密瓜请客人品尝，东道主家中的和谐与温馨，以及全家人的热情和友谊，给德国客人留下了难忘的印象（图 58，D）。2013 年 10 月，吉木萨尔县大龙口村的塔吉克族群众热情地欢迎了波恩大学专家们的到来，并在大龙口剖面点与德国、美国及爱尔兰的专家们合影留念，小朋友们也踊跃参加。一张张珍贵的照片给当地群众和中德专家留下了美好的记忆（图 63，H、I）。

中德新疆合作也像一条条彩带，连接了中德两国与中亚 5 国、俄罗斯和

图 63 中德新疆友谊之花香飘世界

Fig. 63 Flowers of Sino-German friendship fragrant and gorgeous in the world

A. 德国波恩大学、沈阳师范大学及吉林大学联合考察新疆塔里木盆地（2013）；B、D. 美国迪尔切院士参加科研队在新疆的活动；C. 日本科学家西田治文教授（右 1）等参加科研队活动（2007）；E、F. 科研队在德国斯图加特自然史博物馆举办中国带毛恐龙展览（2014）；G. 德国波恩大学师生在中国黑龙江嘉荫野外地质实习（2017）；H、I. 波恩大学来新疆考察的爱尔兰科学家（左 5）等在大龙口与塔吉克族群众合影，H 为大龙口新疆塔吉克族女孩（2013）

A. Sino-German geoscience team organized by Univ. Bonn, SYNU and JU investigating Tarim basin （2013）; B, D. American Prof. Dilcher（NAS）joining the cooperation activities in Xinjiang; C. Japanese scientists headed by Prof. Nishida（right）joining the research team symposium in Xinjiang（2007）; E, F. The research team co-organizing the Special Exhibition on Feathered Dinosaurs of China in Stuttgart Museum（NH）（2014）; G. Students and teachers of Univ. Bonn in Jiayin of Heilongjiang, China for geological excursion（2017）; H, I. Scientists of Univ. Bonn with local Tajik Chinese herdsmen in Dalongkou section: H showing a Tajik little girl（2013）

蒙古等国家的地学合作。受中德新疆地学合作的影响，吉尔吉斯斯坦科学院地质研究所主动要求与中国开展合作研究，2021 年吉尔吉斯斯坦天山地质学会（Tien-Shan Geological Society）主动邀请中、德、俄三国古生物学家参加吉尔吉斯斯坦马德根（Madigen）古生物联合考察。未来，围绕中国新疆的中生代国际合作必将进一步开展。

# 致 谢

本书作者感谢中国科学院－德国马普学会合作项目（1997~1998）、国家自然科学基金重大国际合作项目（30220130698）及合作交流项目（3021133457、3021133458、3021133498、41311130423）、中德科学中心 GZ295 项目、教育部与国家外专局 111 项目（B06008），以及东北亚生物演化与环境教育部重点实验室（吉林大学）、自然资源部东北亚古生物演化重点实验室（沈阳师范大学）、国家古生物化石专家委员会办公室、新疆维吾尔自治区自然资源厅及新疆地矿局、新疆地调院、新疆第一区调大队、新疆维吾尔自治区 305 办公室、吐鲁番地区专署、鄯善县人民政府、中科院南京地质古生物研究所、中科院古脊椎动物与古人类研究所、吉林大学、沈阳师范大学、辽宁古生物博物馆，以及中德合作科研队全体同事的大力支持与热诚帮助。

作者还衷心感谢中国科学院刘嘉麒院士和王永栋教授对本书的推荐，感谢李洁、程显胜等同事在新疆野外工作中的热诚帮助；感谢地质学家沈远超教授对作者前期工作提供的大力帮助；感谢中国科学院李廷栋院士，中德科学中心前主任韩建国、陈乐生，国家自然科学基金委国际合作局前局长卢荣凯，时任新疆地矿局局长田建荣及总工程师董连慧，新疆 305 办公室主任王宝林，新疆第一区调大队领导左学义、宋松山、甄保生，吐鲁番地区专署及鄯善县领导赵文泉、刘开勤、李岩龙，中科院南古所沙金庚、蔡重阳、王成源、尚玉珂、王伟铭、王怿及王鑫甫，南京分院李世杰等专家；感谢中德合作科研队中方专家董枝明、孙春林、孙丰月、吕建生、陈跃军、李春田、王璞珺、刘招君、边伟华、柳蓉、孟庆涛、张建光、李凌及孙巍等，以及德方专家 Pfretzschner H.-U.、Bruch A.、Wings O.、Eder J.、Hinderer M.、Hornung J.、Thein J.、Puettman W.、Eitel B.、Ligouis B.、Joyce W.、Maisch M.、Matzke A.、Uhl D.、Rabi M.、Regent V. 及 Mosbrugger G. 等。衷心感谢俄罗斯地质学家 Pospelov I. 教授在俄罗斯收集 Obruchev 院士来华考察资料，以及

**图 64 中德合作新疆科研队主要成员及协助者们（1997~2020）**

Fig. 64  Main members of Sino-German Co-Research Team and its supporters (1997-2020)

第1排（左起）：孙革，莫斯布鲁格，孙春林，弗莱茨纳，董枝明，阿什拉夫，王璞珺，马丁，孙跃武，李洁；第2排（左起）：泰恩，吕建生，欣德勒，刘招君，皮特曼，王伟铭，怡德，孙丰月，艾特尔，左学义；第3排（左起）：胡侬，陈跃军，布鲁赫，吴文昊，维恩斯，周长付，苗雨雁，麦什，续颜，马茨克；第4排（左起）：乔伊斯，拉比，韦丽娜，边伟华，柳蓉，杨涛，孟庆涛，关谷透，董曼，李凌；第5排（左起）：迪尔切，西田治文，利古伊斯，宋松山，王克卓，甄保生，杜青，李春田，李岩龙，王永栋

Row 1（from left）：Sun G., Mosbrugger V., Sun C. L., Pfretzschner H.-U., Dong Z. M., Ashraf A. R., Wang P. J., Martin T., Sun Y. W., Li J.; Row 2：Thein J., Lv J. S., Hinderer M., Liu Z. J., Puettmann W., Wang W. M., Eder J., Sun F. Y., Eitel B., Zuo X. Y.; Row 3: Hornung J., Chen Y. J., Bruch A., Wu W. H., Wings O., Zhou C. F., Miao Y. Y., Maisch M., Xu Y., Matzke A.; Row 4: Joyce W., Rabi M., Regent V., Bian W. H., Liu R., Yang T., Meng Q. T., Sekiya T., Dong M., Li L.; Row 5: Dilcher D. L., Nishida H., Ligouis B., Song S. S., Wang K. Z., Zhen B. S., Du Q., Li C. T., Li Y. L., Wang Y. D.

日本福井县立恐龙博物馆关谷透博士、中科院徐星及汪筱林教授、山东科技大学周长付教授、沈阳师范大学张宜教授、新华社吉林分社马阳主任等给予的热诚支持与协助（图 64）。

衷心感谢上海科技教育出版社王世平总编、伍慧玲编辑及汤世梁先生、李梦雪女士等对本书的厚爱，以及在编辑出版工作中付出的辛劳和热诚支持。

# 主要参考文献

邓胜徽，程显胜，齐雪峰，等 . 2001. 新疆准噶尔盆地晚三叠世——早侏罗世植物组合序列 . 见：程裕淇等主编 . 第三届全国地层会议论文集 . 北京：地质出版社，174–178.

邓胜徽，王思恩，杨振宇，等 . 2015. 新疆准噶尔盆地中、晚侏罗世多重地层研究 . 地球学报，36（5）：559–574.

董枝明 . 1973. 乌尔禾恐龙化石 . 中国科学院古脊椎动物与古人类研究所甲种专刊，（11）：45–52.

董枝明 . 1977. 吐鲁番盆地的恐龙化石 . 古脊椎动物与古人类 . 15（1）：59–66.

董枝明 . 2009. 亚洲恐龙 . 昆明：云南科技出版社，1–287.

董曼 . 2012. 新疆沙尔湖煤田中侏罗世植物化 . 博士研究生论文 . 长春：吉林大学地球科学学院，1–98.

董曼，孙革 . 2011. 新疆沙尔湖煤田中侏罗世植物化石 . 世界地质，30（4）：497–507.

顾道源 . 1980. 古植物 . 新疆古生物化石图册 . 北京：地质出版社 .

黄嫔 . 2006. 新疆乌鲁木齐郝家沟剖面郝家沟组和八道湾组孢粉组合及其地层意义 . 微体古生物学报，23（3）：235–274.

康玉柱 . 2003. 新疆三大盆地构造特征及油气分布 . 地质力学学报，9（1）：37–47.

李洁，甄保生，孙革 . 1991. 新疆昆仑山乌斯腾塔格——喀拉米兰晚三叠世植物化石的首次发现 . 新疆地质，9（1）：50–58，98–99.

李永安，金小赤，孙东江，等 . 2003. 新疆吉木萨尔大龙口非海相二叠系——三叠系界线层段古地磁特征 . 地质论评，49（5）：525–536.

苗雨雁 . 2003. 疏叶薄果穗（*Leptostrobus laxiflora* Heer）在新疆额敏白杨河中侏罗统的发现 . 吉林大学学报（地球科学版），33（3）：263–269.

苗雨雁 . 2005. 准噶尔盆地中侏罗世西山窑组植物化石新材料 . 古生物学报，44（4）：517–534.（in English）

苗雨雁 . 2006. 新疆准噶尔盆地西部中侏罗世银杏类和茨康类植物 . 硕士研究生论文 . 长春：吉林大学地球科学学院，1–146.

苗雨雁 . 2017. 关于新疆准噶尔盆地中侏罗世奥勃鲁契夫银杏（*Ginkgo obrutschewii* Seward）叶表皮构造的讨论 . 世界地质，36（1）：15–21.

欧阳辉，叶勇 . 2002. 第一具保存完整头骨的马门溪龙——杨氏马门溪龙 . 成都：四川科学技术出版社，1–111.

斯行健，李星学，等 . 1963. 中国植物化石 第二册 中国中生代植物 . 北京：科学出版社，1–429.

孙革 . 1987. 论中国晚三迭世植物地理分区及古植物分区原则 . 地质学报，61（1）：1–9.

孙革.1993.中国吉林天桥岭晚三叠世植物群.长春：吉林科学技术出版社，1–174.

孙革，孟繁松，钱立君，等.1995.三叠纪植物群.见：李星学（主编）.中国地质时期植物群.广州：广东科技出版社，229–259.

孙革，苗雨雁，陈跃军.2006.新疆准噶尔盆地中侏罗世 *Sphenobaiera*（楔拜拉）一新种。吉林大学学报（地球科学版），36（5）：717–722.

唐天福，杨恒仁，兰琇，等.1989.新疆塔里木盆地西部白垩纪至早第三纪海相地层及含油性.北京：科学出版社.

王思恩，高林志.2012.新疆准噶尔盆地侏罗系齐古组凝灰岩 SHRIMP 锆石 U–Pb 年龄.地质通报，31（4）：503–509.

王智，詹家祯，李猛.1999.塔里木盆地东部白垩纪地层划分.新疆石油地质，20（3）.

吴凤鸣.1956.科学家与科学史介绍.著名的苏联地质学家—B. A.奥勃鲁契夫院士.科学通报，7：89，87.

吴绍祖.1990.新疆早侏罗世植物群特征.新疆地质，8（2）：119–132.

吴绍祖，刘兆有.1988.新疆北部白垩纪地层及其沉积特点.新疆地质，6（1）：21–30.

吴文昊，周长付，Wings O，等，2013.新疆鄯善中侏罗世巨型蜥脚类恐龙的发现.世界地质，32（3）：438–462.

新疆区域地质调查大队.1983.中华人民共和国区域地质调查报告（1：20万）白杨河幅（L–45–Ⅷ）.北京：地质出版社，1–256.

新疆地矿局.1993.新疆维吾尔自治区区域地质志.北京：地质出版社，1–841.

席党鹏，万晓樵，李国彪，等.2019.中国白垩纪综合地层和时间框架.中国科学：地球科学，49（1）：257–288.

席党鹏，唐自华，王雪娇，等.2020.塔里木盆地西部白垩纪—古近纪海相地层框架及对重大地质事件的记录.地学前缘，27（6）：165–198.

杨景林，沈一新.2004.准噶尔盆地南缘紫泥泉子组的时空展布及成因解释.地层学杂志，28：215–222.

杨景林，王启飞，卢辉楠.2008.准噶尔盆地白垩纪轮藻化石组合序列.微体古生物学报，24（4）：345–363.

杨景林，沈一新，商华，等.2012.准噶尔盆地南缘露头区紫泥泉子组介形类动物群及时代归属.古生物学报，51（3）：359–369.

杨涛，邢梦蝶，白树崇.2017a.贝加尔茨康叶在新疆白杨河中侏罗统的发现.沈阳师范大学学报（自然科学版），35（4）：288–302.

杨涛，衣李莹，鄂婧文，等.2017b.新疆准噶尔盆地中侏罗世斯卡布勒果（*Scarburgia*）一新种.世界地质，36（2）：327–332.

杨涛，张渝金，鄂婧雯，2018.新疆准噶尔盆地中侏罗统西山窑组费尔干杉属表皮构造研究.地质与资源，27（2）：124–129.

张驰，于兴河，姚宗全，等.2021.准噶尔盆地南缘西段中、上侏罗统沉积演化及控制因素分析.中国地质，48（1）：284–296.

赵喜进.1980.新疆北部中生代脊椎动物化石地层.中国科学院古脊椎动物与古人类所甲种专刊，15：1–120.

周志炎. 1995. 侏罗纪植物群. 见：李星学（主编），中国地质时期植物群. 广州：广东科技出版社，260–309.

周志毅. 2001. 塔里木盆地各纪地层. 北京：科学出版社，1–359.

郑秀亮，郑秀梅，郑新生，等. 2013. 准噶尔盆地晚白垩世介形类化石组合. 地层学杂志，37（2）：206–209.

Ashraf A R, Sun G, Wang X F, et al. 1998. Development of forest-ecosystems and palaeoclimate across the Triassic-Jurassic boundary in the Junggar Basin (NW China)—preliminary results. - The 5th EPPC, June 26-30, 1998 Carcow, Poland ; Wladyslaw Szafer Institute of Botany, Polish Acad Sci, 4.

Ashraf A R, Wang X F, Sun G, et al. 2001. Palynostratigraphic analysis of the Huangshanjie–, Haojiagou– and Badaowan formations in the Junggar Basin (NW China). *Proc. Sino–German Co–Symp. Prehist. Lie Geol. Junggar Bas. Xinjiang, China*, 40–64.

Ashraf A R, Sun Y W, Sun G. 2004. Guide booklet. *In*: Sun G, et al (edit-in-chief). Proceedings of Sino–German Cooperation Symposium on Paleontology, Geological Evolution, and Environmental Changes in Xinjiang, China. Urumqi. Special Material, 1–8.

Ashraf A R, Sun Y W, Sun G, et al. 2010. Triassic and Jurassic palaeoclimate development in the Junggar Basin, Xinjiang, Northwest China—a review and additional lithological data. *Palaeobio Palaeoenv*, 90: 187–201.

Cao R L, Zhu S H, Zhu X K, et al. 1993. Plate and terrain tectonics of northern Xinjiang. *In*: Tu G Z, et al (eds.). New improvement of solid geosciences in northern Xinjaing. Beijing: Sci Press, 11–26.

Chen P J, McKenzie K G, Zhou H Z. 1996. A further research into Late Triassic *Kazacharthra* Fauna from Xinjiang Uygur Autonomous Region, NW China. *Act Palaeont Sin*, 35(3): 272–302.

Cheng Z W, Wu S Z, Fang X S, 1996. The Permian–Triassic sequences in the southern margin of the Junggar Basin and the Turpan Basin, Xinjiang, China. Beijing: Geological Publishing House, 1–25.

Deng S H, Lu Y Z, Fan R, et al. 2010. The Jurassic System of Northern Xinjiang, China. Hefei: Univ Sci Techn China Press, 1–279.

Doludenko M P，Russkazova E S. 1972. 19712 Ginkgoales and Czekanowskiales of the Irkutsk Basin. *In*: Mesozoic plants (Ginkgoales and Czekanowskiales) of East Siberia. Nauka, Moscow, 7–43. (in Russian)

Dong Z M. 1989. On a small Ornithopod (*Gongbusaurus wucaiwanensis* sp. nov.) from Karamaili, Junggar, Xinjiang, China. *Vertebr PalAsiatica*, 27(2):140–146.

Dong Z M. 1990. On remains of Sauropods from Karamaili region, Junggar, Xinjiang, China. *Vertebr PalAsiatica*, 28(1): 43–58.

Dong Z M. 1993. An ankylosaur (Ornithischian dinosaur) from Middle Jurassic of the Junggar, Basin, China. *Vertebr PalAsiatica*, 31(4): 257–266.

Dong Z M. 1997. A gigantic sauropod (*Hudiesaurus sinojaponorum* gen. et sp. nov.) from the

Turpan Basin, China. *In*: Dong Z M (ed). *Sino–Japanese Silk Road Dinosaur Expedition*. Beijing: China Ocean Press, 102–110.

Dong Z M. 2001. Mesozoic fossil vertebrates from the Junggar Basin and Turpan Basin, Xinjiang, China. *Proc. Sino–German Co–Symp. Prehist. Lif. Geol. Junggar Bas. Xinjiang, China*: 95–103.

Florin R. 1936. Die fossilen Ginkgophyten von Franz–Joseph–Land nebst Erörterungen über vermeintliche Cordaitales mesozoischen Alters. I. *Palaeontographica* B, 81: 71–173.

Halle A. 1908. Zur Kenntnis der mesozoischen Equisetales Schwedens. *K. Svensk. Vet. Akad. Handl.*, 7: 113.

Harris T M. 1951.The fructification of *Czekanowskia* and its allies. *Philophical Transactions of the Royal Society of London Series*, 235: 483–508.

Harris T M. 1961. The Yorkshire Jurassic Flora, I. *Thallophyte–Pteridophyta*. London: British Museum (Nat Hist), 1–204.

Harris T M, Millington W. 1974. The Yorkshire Jurassic flora. IV .1. Ginkgoales; 2. Czekanowskiales. London: British Museum of London, 2–150.

Hornung J, Sun G, Li J, et al. 2003. Fluvial–deltaic lithofacies and architectural element analyses at the Haojiagou–valley section in the Junggar–Basin (NW–China, Middle/Upper Triassic). *Palaeontographica*, in review.

Hsu R, et al. 1979. *Late Triassic flora of Baoding, China*. Beijing: Sci Press, 1–130.

Hu A Q, Zhang G X, Li Q X, et al. 1997. Isotopie geochemistry and crustal evolution of northern Xinjiang. *In* Tu G Z, et al. (eds). *New improvement of solid geosciences in northern Xinjiaing*. Beijing: Sci Press, 27–37.

Hu Y M, Meng J, Clark J M. 2007. A new Late Jurassic docodont (Mammalia) from northeastern Xinjiang, China. *Vertebr PalAsiatica*, 45: 173–194.

Huang Z G, Zhou H Q. 1980. Plants. *In*: Stratigraphy and palaeontology of the Shaanxi, Gansu and Ningxia Basin. Beijing: Geological Publishing House, 185. (in Chinese)

Kiritchkova A I, Travina T A, Bystritskaya L I. 2002. The *Phoenicopsis* Genus: Systematics, History, Distribution and Stratigraphic Significance. Petersburg: VNIGRI, 1–205. (in Russian)

Krassilov V A. 1968. On the study of fossil plants of Czekanowskiales. *Transactions of Geological Institute, Academy of Science USSR*, 191: 31–40. (in Russian)

Krassilov V A. 1972. Morphology and systematics of Ginkgoles and Czekanowskiales. *Paleontological Journal*, 1: 113–118. (in Russian)

Li C Y, Wang Q, Liu X Y, et al. 1982. *Explanatory notes to the tectonic map of Asia*. Beijing: Cartographic Press.

Li P J, He Y L, Wu X W, et al. 1988. *Early and Middle Jurassic strata and their floras from northeastern bounder of Qaidam Basin, Qinghai*. Nanjing: Najing Univ Press, 1–231.

Ligouis B. 2001. Organic petrography, geochemistry, stratigraphy, facies geology and basin analysis in the Triassic and Jurassic of the Junggar–Basin (NW China). *Proc. Sino–German*

*Co–Symp. Prehist. Lif. Geol. Junggar Bas. Xinjiang, China*, 104–113.

Maisch M W, Matzke A T, Pfretzschner H–U, et al. 2001. The fossil vertebrate faunas of the Toutunhe and Qigu formations of the southern Junggar Basin and their biostratigraphical and paleoecological implications. *Proc. Sino–German Co–Symp. Prehist. Lif. Geol. Junggar Bas. Xinjiang, China*: 83–94.

Maisch M W, Matzke A T, Sun G. 2004. A relict trematosauroid (Amphibia: Temnospondyli) from the Middle Jurassic of the Junggar Basin (NW China). *Naturwissenschaften*, 91 (12): 589–593.

Maisch M, Matzke A, Grossmann F, et al. 2005. The first haramiyid mammal from Asia. *Naturwissenschaften*, 92: 40–44.

Martin T. 2004. Incisor enamel microstructure of South America's earliest rodents: implications for caviomorph origin and diversification. *In*: The Paleogene Mammalian Fauna of Santa Rosa, Amazonian Peru. K. E. Campbell, Jr., editor, Natural History Museum of Los Angeles County, *Science Series*, 40: 131–140.

Martin T, Averianov A O, Pfretzschner H–U. 2010a. Mammals from the Late Jurassic Qigu Formation in the Southern Junggar Basin, Xinjiang, Northwest China. *Palaeodiv Palaeoenv*, 90: 295–317.

Martin T, Sun G, Mosbrugger V. 2010b. Triassic–Jurassic biodiversity, ecosystem, and climate in the Junggar Basin, Xinjiang, Northwest China. *Palaeobio Palaeoenv*, 90: 171–173.

Matzke A T, Maisch M W, Sun G, et al. 2004. A new Xinjiangchelyid turtle (Testudines, Eucryptodira) from the Jurassic Qigu Formation of the southern Jungar Basin, Xinjiang, North–West China. *Palaeontology*, 47:1267–1299.

Matzke A T, Maisch M W, Sun G, et al. 2005. A new Middle Jurassic xinjiangchelyid turtle (Testudines; Eucryptodira) from China (Xinjiang, Junggar Basin. *J Vertebr Paleont*, 25: 63–70.

Nebelsick J. 2004. Carbonatic facies dynamics during period of global climatic transition. *Proc. Sino–Germ. Symp. Paleont. Geol. Evul. Env. Chang., Urumqi*, 73–75.

Nosova N, Zhang J W, Li C S. 2011. Revision of *Ginkgoites obrutschewii* (Seward) Seward (Ginkgoales) and the new material from the Jurassic of Northwestern China. *Review of Palaeobotany and Palynology*, 166: 286–294.

Obruchev V A. 1914. Diam River, Orkhu Basin and Mt. Khara–Arati. Border area of Dzungaria. Route Investigation. *Bulletin of Tomsk Technological Institute, Russia*, 1 (2): 433–437.

Obruchev V A. 1940a. Geological map of boundary region of Dzungaria (1:500000 in scale). *In* Obrutschev V A. Boundary region of Dzungaria. M–L: Acad Sci USSR, 3.

Obruchev V A. 1940b. Regional Dzunggaria. Geology. Geographical and Geological Records. Geological Science Institute, Acad Sci USSR. M–L: Acad Sci USSR, 3 (2): 138–156.

Pfretzschner H–U, Ashraf A R, Maisch M, et al. 2001. Cyclic growth in dinosaur bones from the Upper Jurassic of NW China and its paleoclimatic implications. *Proc. Sino–German Co–*

*Symp. Prehist. Lif. Geol. Junggar Bas. Xinjiang, China*, 21–39.

Pfretzschner H–U, Martin T, Maisch M, et al. 2004. A new docodent from the Totunhe Formation (Middle Jurassic) of the Junggar Basin. *Proc. Sino–Germ. Symp. Paleont. Geol. Evul. Env. Chang.*, Urumqi: 9–11.

Pfretzschner H–U, Martin T, Maisch M, et al. 2005. A new docodont mammal from the Late Jurassic of the Junggar Basin in Northwest China. *Acta Palaeontol Pol*, 50(4): 799–808.

Rabi M, Zhou C F, Wings O, et al. 2013. A new xinjiangchelyid turtle from the Middle Jurassic of Xinjiang, China and the evolution of the basipterygoid process in Mesozoic turtles. *BMC Evolutionary Biology*, 13: 203.

Regional Geological Survey of Xinjiang. 1965. Geology of Urumqi areas (K–45–IV). Beijing Print Fact, 8: 1–35.

Regional Geological Survey of Xinjiang (RGSX), Geological Institute of Xinjiang, Geological Surveying Division of Petroleum Bureau of Xinjiang. 1983. *Palaeontological Atlas of Northwest China*. Vol. Xinjiang. Beijing: Geological Publishing House, 1–795.

Samylina V A. 1967. On the final stages of the history of the genus *Ginkgo* L. in Eurasia. Z. Bot. 52: 303–316.

Samylina V A, Kiritchkova A I. 1991. The genus *Czekanowskia* (Systematics, History, Distribution and Stratigraphic Significance). Leningrad: Nauka, 1–130. (in Russian)

Samylina V A, Kiritchkova A I. 1993. The genus *Czekanowskia* Heer. Principles of systimatics, range in space and time. *Review of Palaeobotany and Palynology*, 79: 271–284.

Seward A C. 1911. Jurassic plants from Chinese Dzungaria, collected by Professor Obrutschew. *Mem. Com. Geol. N.S. St. Petersbourg*, 75: 1–61.

Sun A L. 1978. Two new genera of Dicynodontidae. In Permian and Triassic vertebrate fossils of Dzungaria basin and Tertiary stratigraphy and mammalian fossils of Turfan basin. *Mem Inst Vertebr Paleont Paleoanthropol, Acad Sin*, 13:19–25.

Sun C L, Dilcher D L, Wang H S, et al. 2009. *Czekanowskia* from the Jurassic of Inner Mongolia, China. *International Journal of Plant Sciences*, 170 (9): 1183–1194.

Sun C L, Na Y L, Dilcher D L, et al. 2015. A new species of *Phoenicopsis* subgenus *Windwardia* (Florin) Samylina (Czekanowskiales) from the Middle Jurassic of Inner Mongolia, China. *Acta Geologica Sinica*, 89(1): 55–69.

Sun G. 1979. On the discovery of *Cycadocarpidium* from the Upper Triassic of eastern Jilin. *Act Palaeont Sin*, 18 (3): 312–326.

Sun Ge. 1993. *Ginkgo coriacea* Florin from Lower Cretaceous of Huolinhe, northeastern Nei Monggol, China. *Palaeontographica* B: 159–168.

Sun G. Meng F S, Qian L J, et al. 1995. Triassic flora of China. *In*: Li X X (eds–in–chief). *Fossil plants of China through the geological ages*. Guangzhou: Guangdong Sci Techn Press, 229–259.

Sun G, Mosbrugger V, Ashraf A R, et al. 2001a. The advanced study of prehistory life and geology of Junggar Basin, Xinjiang, China. Urumqi, 1–113.

Sun G, Mosbrugger V, Li J, et al. 2001b. Late Triassic flora from the Junggar Basin，Xinjiang, China. *Proc. Sino−German Co−Symp. Prehist. Lif. Geol. Junggar Bas. Xinjiang, China*, 8–20.

Sun G, Miao Y Y, Sun Y W, et al. 2004a. Middle Jurassic plants from Baiyang River area of Emin, northwestern Junggar Basin, Xinjiang. *In*: Sun G, et al(eds). Proc. Sino−German. Symp. Geol. Evol. Environm. Changes of Xinjiang, China. *Urumqi*: 35–40.

Sun G, Mosbrugger V, Ashraf A R, et al. 2004b. Paleontology, geological evolution and environmental changes of Xinjiang, China. *Proc. Sino−Germ. Symp. Paleont. Geol. Evol. Env. Chang., Urumqi*: 1–96.

Sun G, Miao Y Y, Chen Y J. 2006. A New species of Sphenobaiera from Middle Jurassic of Junggar Basin, Xinjiang, China. *Jilin Univ (Earth Sci Edit)*, 36 (5): 717–722.

Sun G, Miao Y Y, Mosbrugger V, et al. 2010. The Upper Triassic to Middle Jurassic strata and floras of the Junggar Basin, Xinjiang, Northwest China. *Palaeobio Palaeoenv*, 90: 203–214.

Sun G, Yang T, Tan X, et al. 2021. Recent advances in the study of the ferns from Middle Jurassic Baiyanghe flora of Junggar Basin, Xinjiang, China. *Palaeobio Palaeoenv*, 101: 19–24.

Sze H C. 1933. Fossile Pflanzen aus Shensi, Szechuan und Kuuichow. *Palaeont Sin*, Ser A, 1 (3): 1–32.

Taylor T N, Taylor E L. 2009. The Biology and Evolution of Fossil Plants. Englewood Cliffs: Prentice Hall, 744–787.

Vakhrameev V A. 1988. Jurassic and Cretaceous floras and climates of the Earth. Nauka: 1–210. (in Russian)

Vakhrameev V A, Doludenko M P. 1961. Late Jurassic and early Cretaceaus flora from Bureja Basin and its significance for stratigraphy. *Trans Geol Inst Acad Sci USSR*, 54: 1–135.

Wang T, et al. 1996. Oil and gas reservoir geology of rift basins in China. Beijing: Petroleum Industry Press, 1–172.

Wang X L, Kellner A W, Jiang S, et al. 2014. Sexually dimorphic tridimensionally preserved pterosaurs and their eggs from China. *Current Biology*, 24 (12): 1323–1330.

Wang X L, Kellner A W, Cheng X, et al. 2017. Egg accumulation with 3D embryos provides insight into the life history of a pterosaur. *Science*, 358(6367): 1197–1201.

Wang Y D, Zhang W, Saiki K. 2000. Fossil wood from the Upper Jurassic of Qitai, Junggar Basin, Xinjiang, China. *Acta Paleontologica Sinica*, 39 (Suppl.): 176–185.

Wang Y D, Shi Y, Miao Y Y. 2002. New advances of palaeontology and geology in the Junggar Basin, Xinjiang. *Episodes*, 25 (2): 128–131.

Wings O, Schellhorn R, Mallison H, et al. 2007. The first dinosaur tracksite from Xinjiang, NW China (Middle Jurassic Sanjianfang Formation, Turpan Basin)—a preliminary report. *Global Geology*, 10(2): 113–129.

Wings O, Rabi M, Schneider J W, et al. 2012. An enormous Jurassic turtle bone bed from the Turpan Basin of Xinjiang, China. *Naturwissenschaften*, 99 (11): 925–935.

Wu C Y, Xue S H, et al. 1997. Sedimentology of petroliferous basins in China. Beijing: Petroleum Industry Press, 401–411.

Wu S Q, Zhou H Z. 1986. Early Jurassic plants from East Tianshan Mountain. *Act Palaeont Sin*, 25 (6): 636–647.

Wu S Z, Li Y A, Qu X, et al. 2001. Triassic geography and climate of the Junggar Basin. *In*: Sun G, et al (eds). Proc. Sino–German Symp. Prehist. Lif. Geol. Junggar Basin, Xinjiang, China, 1–9.

Xing L D, Klein H, Lockley M G, et al. 2014. *Changpeipus* (theropod) tracks from the Middle Jurassic of the Turpan Basin, Xinjiang, Northwest China: review, new discoveries, ichnotaxonomy, preservation and paleoecology. Vertebrata PalAsiatica, 52(2): 233–259.

Xu X, Clark J M, Forster C A, et al. 2006. A basal tyrannosauroid dinosaur from the Late Jurassic of China. *Nature*, 439: 715–718.

Yang J D, Qu L F, Zhou H Q, et al. 1986. Permian and Triassic strata and fossil assemblages in the Dalongkou area of Jimsar, Xinjiang. *Geol. Mem*. Ser. 2, No. 3. Beijing: Geological Publishing House, 1–262.

Young C C. 1935. On two skeletons of Dicynodontia from Sinkiang. *Bull Geol Soc China*, 14: 483–517.

Young C C. 1978. A Late Triassic vertebrate fauna from Fukang, Sinkiang. *Mem Inst Vertebr Paleont Paleoanthropol, Acad Sin*, 13: 60–67.

Yuan P L, Young C C. 1934. On the discovery of new *Dicynodon* from Sinkiang. *Bull Geol Soc China*, 13: 563–574.

Zhang X Q, Li D–Q, Xie Y, et al. 2018. Redescription of the cervical vertebrae of the mamenchisaurid sauropod *Xinjiangtitan shanshanesis* Wu et al. 2013. *Historical Biology*, 32 (5): 1–20.

Zhao X J. 1980. Mesozoic vertebrate fossils and strata from the northern Xinjiang. *Mem Inst Vertebr Paleont Paleoanthropol, Acad Sin*, 15: 1–120.

Zhen B S, Li J, Sun G. 1991. First discovery of the Late Triassic from Wustentag–Haramilan of Kunlun Mt., Xinjiang. Geol. Xinjaing, 9: 1–6.

Zhou C F, Bhullar B A S, Neander A I, et al. 2019. New Jurassic mammaliaform sheds light on early evolution of mammal–like hyoid bones. *Science*, 365: 276–279.

Zhou Z Y. 1995. Jurassic flora. *In*: Li X X (eds–in–chief). *Fossil floras of China through the geological ages*. Guangzhou: Guangdong Sci Techn Press, 343–410.

Zhou Z, Zhan B L. 1989. A Middle Jurassic Ginkgo with ovule–bearing organs from Henan, China. *Palaeontographica Abt B*, 211: 113–133.

Zhou Z Y, Lin H L. 1995. Stratigraphy, paleogeography and plate tectonics from Northwest China. Nanjing: Nanjing Univ Press, 1–299. (in Chinese)

Zhou Z Y, Dean W T. 1996. Phanerozoic geology of Northwest China. Beijing: Sci Press, 1–316.

# Mesozoic of Xinjiang, China

by

Sun G.,  Mosbrugger V.,  Sun Y. W.,  Ashraf A. R.,  Martin T.,

Miao Y. Y., Wang K. Z.,  Wu W. H.,  Yang T.,  Dong M.

# Preface

Xinjiang, the biggest administrative province of China, is located in northwestern China, covering an area of over 1.66 million km$^2$, about 1/6 of Chinas total land territory. Xinjiang, well-known for the Silk Road, links Central Asian countries, Russia, and western Mongolia. In the Mesozoic Era (ca. 252-66 million years ago), rich oil, coal, and natural gas deposits developed in the Junggar-, Tarim-and Turpan-Hami (Tu-Ha) basins of Xinjiang, which are very beneficial to the economic development of China. Looking at the Mesozoic strata of Xinjiang is like flipping through a geological textbook: taking a broad view down from the top of the Tianshan Mountains (mostly Carboniferous strata of over 300 million years), the undulate low mountains and hills are mostly composed of Mesozoic strata, which are colorful and overlapping, just like exquisite pictures (Fig. 1).

The three basins of Xinjiang have yielded abundant Mesozoic fossils, including tetrapsid, dinosaurs, pterosaurs, turtles, crocodiles, sharks, fish, mammals, invertebrates, and plants, thus providing a "natural laboratory" to study the

Mesozoic biological evolution and environmental changes, as well as valuable basic data for searching fossil fuels such as oil, natural gas, and coal in Xinjiang. Therefore, the unique paleontological resources in Xinjiang and especially its Mesozoic fossils have recently attracted more attention from geoscientists world-wide.

Xinjiang has a long history in geoscience research, which can be traced back to the Russian geologist Obruchev V. A. on 1904-1909 or even earlier. Chinese geologists Yuan F. L., Huang J. Q., and Cheng Y. Q. also did a lot of research work between 1930s and 1940s. After the founding of the People's Republic of China (in 1949), the former Xinjiang Bureau of Geology and Mineral Resources (BGMRX), Institution of Petroleum, Xinjiang (IPX) and Bureau of Coal Geology, Xinjiang (BCGX), Chinese Academy of Sciences (CAS), Chinese Academy of Geosciences (CAGS) and other institutions, have carried out a lot of research work on the Mesozoic geology and paleontology in Xinjiang (BGMRX, 1993). The above achievements have laid a valuable foundation for the following research.

Viewed as a whole today, mainly due to its inland location, and being far away from the sea and in the hot and dry climates, about 60% of Xinjiang is covered by Gobi and desert. However, the Mesozoic Xinjiang was a different scene: about 252 million years (Ma) ago, Xinjiang was covered by seawater, and then mainly uplifted to land, although some areas still remained sea-bays in southwestern Tarim Basin around 100 Ma ago. During the Mesozoic, the climate in Xinjiang was warm and humid and hence many rivers and lakes existed in the Junggar Basin, Tarim Basin, and Tu-Ha Basin, where fish, turtles, crocodiles, bivalves, gastropods, conchostracans, ostracods, and other animals lived. The mammal-like reptiles (e.g. *Lystrosaurus*, etc.), dinosaurs (e.g. *Xinjiangotitan*, etc.), and early mammals (e.g. *Dsungarodon*, *Sineleutherus*, etc.) lived on the shore; the pterosaurs (e.g. *Dsungaripterus*, *Hamipterus*, etc.) soared in the air; and the lush green forests on the earth created fresh environment here, showing a scene full of vitality. However, since the Cenozoic Era, along with great geological events such as the uplift of the Qinghai-Tibet Plateau, the forming of the East Asian monsoon, and the change of direction of the Western monsoon, the terrestrial ecosystems of Xinjiang experienced great changes, resulting in the forming of the endless Gobi desert, and the "biological paradise" gradually disappeared in Xinjiang. Thus, to better understand the bio-/geological evolutionary history of the Mesozoic Xinjiang, particularly of its terrestrial ecosystem changes, which are also directly related to the exploration of the fossil energy resources as mentioned above, the Sino-German Cooperative Geoscientific Research Team has made intensive research on the Mesozoic biotas and strata in Xinjiang for about 20 years. This book has been written to inform an international readership of this research.

The book consists of five chapters with 21 sections. Chapters 1, 4, and 5 mainly focus on the stories about the Sino-German Research Team led by Prof. Dr. Sun G. and Prof. Dr. Mosbrugger V., and their research done in Xinjiang since 1997. Chapters 2 and 3 mainly highlight their professional achievements in the Mesozoic biostratigraphy and fossils in Xinjiang. The achievements include many interesting discoveries, such as the new findings of the largest known Jurassic dinosaur *Xinjiangotitan*, the dinosaur tracks in Shanshan, the early mammal new taxa *Dsungarodon* and *Sineleutherus*, the new species of Jurassic turtles *Xinjiangchyles*, the Early Triassic dicynodontid therapsid *Lystrosaurus*, the Late Triassic Haojiagou flora, and a new round of studies of the Middle Baiyanghe flora and Shaerhu flora, etc. Some stories, e. g. "tracing Obruchev's road" , "three round working in Shanshan Gobi desert", "emotion with fossils in Xinjiang", "sound of dinosaurs in Turpan" and "the second life", and the appearance of the selected "Top Ten Fossil Stars of Mesozoic Xinjiang" may arouse readers' particular interest. All in all, this book is the first Sino-German cooperative book on geoscience that is both scientific and popular. It is hoped that readers can get not only more scientific knowledge but also more impressions about the scientific spirit of Chinese and German scientists.

The authors would like to express their sincere thanks for the support for the Program of CAS (China)-Max Planck Society (Germany) in 1997 and 1998, the projects of NSFC (China), DFG (Germany) since 1999, and the Sino-German Science Promotion Center (GZ295); Project 111 (B06008) and Key-Lab of Evolution of Past Life and Environment of NE Asia of Ministry of Education, China; the Key Lab of Evolution of Past Life of NE Asia, Ministry of Natural Resources, China; the NIGPAS, IVPP, JU, SYNU, and PMOL of China; the Office 305 of Xinjiang; the Geological Bureau and Survey of Xinjiang, RGSX No.1; the local governments of Turpan Region including the Shanshan County; and all members of the Sino-German Research Team in Xinjiang. Without their kind support and assistance, the work could not have been carried out so successfully.

Sincere thanks also go to Academician Prof. Liu J. Q. (IGG CAS) and Prof. Wang Y. D. (NIGPAS) for their kind recommendation of this book for publication; to Academician Li T. D. for his warm support for this cooperative work; to Prof. Shen Y. C. (IGG CAS) and Dr. Cheng X. S. for their great help to the authors' early work for the project; to Dr. Pospelov I. (IG RAS) for his kind help in finding and providing the important data of Academician Prof. V. Obruchev, housed in the VSEGEI, Russia.

*Mesozoic Xinjiang, China* is a comprehensive work that brings together the diverse expertise and broad knowledge of 20 years of geoscientific research and Sino-German cooperation and friendship in the beautiful, but harsh environment of Xinjiang. It is hoped that it may encourage more research and scientific publications in the coming future.

October, 2021

# Chapter ① 

# Tracing Obruchev's road

## 1.1 Tracing the expedition road of Obruchev

### 1.1.1 Expedition of Obruchev in Junggar of Xinjiang, China

Academician Prof. Obruchev V. A. (1863-1956), one of the most outstanding geologists in Russia, graduated from the St. Petersburg Institute of Mining Technology in 1886. He has been Professor at the Tomsk Institute of Technology since 1901, Professor at the Moscow Institute of Mining Technology (1921-1929), and the Director of the Frozen Soil Research Institute of the Academy of Sciences, USSR since 1930. He was elected Member of the AS USSR in 1929, and won the Lenin Medal several times and the first Karbinsky Gold Medal. He was engaged in the geological research of Siberia, Mongolia, and Central Asia for a long time. Between 1904 and 1905 and in 1909, he made two expeditions to the Junggar Basin of China via Mongolia, collected a lot of fossils, and drew the first geological map of the Junggar region (1 : 500000) (Obruchev, 1914, 1940a) (Figs. 2 ; 65).

In 1904-1905, Obruchev collected many plant fossils in the Ak-djar ravine of the Diam River in the northwestern Junggar Basin, and investigated its geology and geography. After his return to Russia, he sent the fossils to Prof. A. C. Seward (FRS), the famous paleobotanist of the British Museum (Natural History) for identification; Seward identified 19 species of 16 genera with two new species, and considered the age of the plant fossils to be Jurassic. After that, Seward published the well-known paper "Jurassic plants in Junggar, China" (Seward, 1911), and Obruchev also used the fossil identifications in his book *On Geography and Geology of Junggar Region* (Obruchev, 1914, 1940b). Both experts pointed out that the fossils were found in the Ak-djar ravine of the "Diam River" in their books, and used the name "diam" of the original fossil site for the new species of the fern *Raphaelia diamensis* Seward in nomenclature. The Ak-djar ravine with the Diam River were both clearly marked in Obrutchev's Geological Map of the Junggar region (1 : 500000) (Figs. 2, C; 65, C). However, unfortunately, both "Diam" and "Ak-djar" sites are not detectable on the maps of China. Later, due to multiple reasons including the diplomatic relations between China and the USSR, it was difficult to find the geological map of Obruchev. Though more than half a century passed, still no one knew where the "Diam" and "Ak-djar" were, and even no one was interested in this question. Thus, the "Diam and Ak-djar" remained mysteries in China, at least for the Chinese geologists and paleontologists, for a long time.

## 1.1.2 Solving the mystery of "Diam"

In 1988 and 1989, Dr. Sun G. was granted the Sino-British Friendship Scholarship Scheme (SBFSS) to go to the British Museum (Natural History) in London as a visiting scholar with a postdoctoral level. During the time he worked in the "Seward Library" for a month. As a paleobotanist, Sun G. showed great respect for the outstanding paleobotanist, Prof. A. C. Seward (FRS), and some familiar with his contributions to the study of Jurassic plants in Junggar, China (Seward, 1911), even frequently used the specific name "*Raphaelia diamensis* Seward" in the identification of

Chinese fossils. Besides the honor of working in Seward's previous office, Sun G. also increased his interest in Seward's research on the Jurassic plants in Junggar of Xinjiang, China. However, after checking the collection records, Sun G. realized that all the plant fossils from the Junggar were returned to Obruchev, when Seward had finished examination of these fossils, and later were housed in the Russian Geological Museum (VSEGEI) in St. Petersburg. This made it even more difficult to find out the

Hoboksar

Baiyang River valley and geological section

Fig. 65  Tracing the fossil site Baiyanghe (Diam River) of Obruchev

A. Russian academician Prof. Obruchev V. A.(1863-1956); B. Geological expedition route of Obruchev from Mongolia to Junggar of Xinjiang, China (1904-1905); C. Geological map of Junggar (B and C after Obruchev, 1914, 1940a); D. Route for searching of the fossil site of Obruchev by Sun G. et al. in 1989; E. Hoboksar ; F, G. Baiyang (Diam) River valley and its geological section

Baiyang (Diam) River Tiechanggou

N

G

locations of "Diam" and "Ak-djar". Sun G. kept this in his mind, and he decided to find out the two localities where these fossils were found after his return to China, which would be helpful for the further study of Jurassic plants in Junggar of Xinjiang, and also for solving the "mystery of Diam" as soon as possible.

In February 1989, Sun G. returned from London to Nanjing Institute of Geology and Paleontology (NIGPAS) , China. Not long after, he was invited by Prof. Shen Y. C., a close friend and famous geologist working in the Institute of Geology, Chinese Academy of Sciences (IG CAS) in Beijing who was taking field work in Xinjiang. Prof. Shen was the chief-scientist of a project of the Office 305, Xinjiang, for prospecting gold deposits at that time. Taking this opportunity, Sun G. flew to Urumqi in June of 1989, and then went to the Regional Geological Survey No. 1 of Xinjiang (RGSX, No. 1) in Qitai, to ask Ms. Li J., a paleobotanist of the Survey for assistance. Li J. was Sun's young colleague when she was a visiting scholar in the NIGPAS for advanced study, and got the help from Sun G.

Thus, everybody got ready, and Sun G. and Li J. with a "Beijing jeep" for field use, and a driver, provided by Prof. Shen, started to trace the expedition road of Obruchev in the Junggar of Xinjiang, from Karamay, the famous "oil city" (Fig. 65).

At first, Sun, Li, and the driver arrived in Hoxtolgay, a coal-mine town in northwestern Junggar, Xinjiang. They collected some fossil plants, but with no information on the places "Diam" and "Ak-djar". Two days later, they arrived in the Hoboksar (Hefeng County), the border county between China and Russia, in the Junggar Basin, where they learned from a Kazakh Chinese staff of the local government that "Diam" means "Baiyang" in the Kazakh language, and that there is a place "Baiyang River" near the Tiechanggou, a small coal-mine town, about 200 km from Hoboksar Town. This was great news for Sun G. and his companions, and the places were consistent with the geological map Sun had. Too excited, they did not care about the beautiful scenery of the boundless grassland of Hoboksar in the embrace of mountains, and headed for Tiechanggou, the small coal mine town, to find the "Baiyang River", on the early morning of the following day.

## 1.1.3 Finding the place of Baiyang River with fossil sites

Tiechanggou is a small coal mine town located about 200 km southwest of Hoboksar (Hefeng) County. However, the two places are separated by the Semistai Mt., which can be crossed through a road pass where used to be an abandoned rare metal mine. The road of the mountain pass is old and rarely used, and past the pass, there is the endless Gobi with almost no road available to Tiechanggou town. As experienced geologists, Sun G. and his two colleagues were confident to overcome the difficulty of the road, but a mere 200 km distance took them nearly ten hours. It was already afternoon when they arrived at the abandoned rare metal mine near the pass, and after going down the mountain, they saw the vast Gobi with no road for driving. But in the distance, they could discern a faint shadow of Tiechanggou town. Sun G. and his two companions regained confidence, because the target in the distance was consistent with the direction of the map. However, the remaining 60-70 km distance was on bumpy "washboard road" which heavily shook the car, and that was the most difficult time for traveling in the Gobi. The "washboard" was caused by ravines from the hillside cutting the Gobi. In any case, the victory finally belongs to the brave people, and at dusk, Sun and his companions finally arrived at the Tiechanggou coal mine and found accommodation in a simple guest house.

The next morning, according to the direction shown on the map, Sun G. and his companions drove eastward from Tiechanggou, following the fuzzy skid marks or the footprints of sheep on the Gobi, and after more than an hour's journey, they sighted, a north-south extending terrace. At the top of the terrace, overlooking the bottom of the valley, they had a good view of a dense poplar forest, like a broad green belt, in which a winding river ran through the forest from north to south. On its eastern side, a perfect coal-bearing stratigraphic section was cut by the river valley. Oh, this was the Baiyang River, which they dreamed for a long time, and finally found! Checking Obruchev's geological map, they became aware that this valley is exactly the "Ak-djar

Ravine" of Obruchev, hiding the mysterious section and coal-bearing site. From the conversation late on with local Kazakh herdsmen, Sun G. learned that in the Kazakh language, the original "Ak" means "white", while the "djar" means "tree", and together, "Ak-djar" means "white trees", which possibly refers to the poplars that are growing there. Thus the river later was named "Diam River", and then Baiyang River ("white poplar river" in Chinese) (Fig. 65, F, G).

Walking into the valley of the Baiyang River, you can be greeted by many abandoned earth walls that are easily seen. An old Kazakh herder said that there used to be coal diggers and herdsmen living at the site. The old man also told Sun G. that he lived at ease, with a good income from selling about 100 sheep he owned, which enabled him to employ a young man for keeping the herds. He appreciated the government's policy, and enjoyed himself with a drink every day, without worrying about food and clothing (Fig. 4, C).

The Baiyang River valley section (i.e. "Ak-djar section") is a very spectacular sight, about 3 km long from north to south. The Jurassic coal-bearing strata are mainly composed of gray sandstone and siltstone, with at least 3-4 thick coal seams. The fossil plants here are almost visible from above. The perfect geological profile and wonderful natural landscape are very impressive. Due to the limited time allowed for the use of the car, Sun G. and his colleagues only worked here for two days, collected nearly 100 well-preserved plant fossils in four wooden boxes, and made a preliminary survey of the section, which was quite fruitful. At this point, the "mystery of Diam" was finally solved, and hence Sun G. and his companions have been closely tied to Xinjiang, especially the Baiyang River.

# 1.2  Imaginations on cooperation initiated in Bonn and Tuebingen

## 1.2.1  Imaginations on cooperation in Bonn and Tuebingen

The friendship between Dr. Sun G. (China) and Dr. Mosbrugger V. (Germany) was established in 1986 when Mosbrugger was invited to visit Nanjing, China. Mosbrugger was a postdoc. in paleobotany at the time, working in the Institute of Paleontology, University of Bonn, Germany, and he had received a research task from his supervisor, Prof. Schweitzer H.-J., the world-famous paleobotanist on the study of the Permian plants from Baode of Shanxi, China. These fossils were collected from Baode in the 1920s by the famous Swedish paleobotanist Prof. Halle T. who did not finish their study. Thus, the Swedish Museum of Natural History in Stockholm where Prof. Halle was working and where the specimens are housed, asked Prof. Schweitzer to continue this work, which finally was entrusted to Mosbrugger, who gladly accepted this task, and was full of enthusiasm when he traveled to Nanjing, China. He was kindly invited to make this visit by the Chinese Academician Prof. Li X. X., the famous paleobotanist of the NIGPAS, who was also an old friend of Prof. Schweitzer. In addition, Mosbrugger has an old friend, Prof. Cai C. Y. in the NIGPAS, who was a visiting scholar with the German "Humboldt Scholarship" working in Bonn for two years. Since Sun G. was the Ph. D. researcher supervised by Prof. Li, and a junior fellow apprentice and a close friend of Cai, both were very pleased to receive Mosbrugger for his visit to China. Thus, for preparation of the field work for Mosbrugger, Sun G. gladly undertook a field trip to Baode of Shanxi for a week. And later, Cai was accompanying Mosbrugger during the field trip in Baode. During the visit of Mosbrugger to Nanjing, he not only examined the plant fossils kept in the NIGPAS, but also gave a wonderful lecture to the Chinese colleagues, which was highly appreciated. Thus, through the friendly reception and communication in Nanjing, Sun G. became a good friend with Mosbrugger also. During the time, besides the scientific exchange, Mosbrugger was also invited to visit Sun G.'s home. Since then, they have developed a deep friendship with each other (Fig. 5).

In 1988, while Sun G. was the visiting scholar of the British Museum (NH) in London, he was kindly invited by Prof. Schweitzer to visit the Institute of Paleontology of the University of Bonn, Germany in September. As old friends, Sun G. and Mosbrugger were very happy to see each other after a long time. In Bonn, they analyzed together with their colleague, palynologist Dr. Ashraf A. R., the plant fossils from the Huolinhe Coal Mine of China, brought along by Sun G. They found very significant fossil cuticles and sporopollen well preserved in the Chinese fossils. Thus, the idea of cooperative research on the Chinese Huolinhe coal-bearing strata and flora came up between the two sides. Soon, a Sino-German cooperative application was sent to the Volkswagen Foundation of Germany by Sun G. from the Chinese side. Unfortunately due to a reduction in the scope of the Volkswagen program, the application was declined. Thus, Sun G. and Mosbrugger waited for an opportunity in the future for cooperative research to be carried out (Fig. 6).

In 1990, Dr. Mosbrugger was appointed Professor by the University of Tuebingen (UT), and later became the Dean of the Institute of Geosciences, UT; while Sun G. was promoted to professorship in 1994 and Deputy Director of the NIGPAS in 1995, which created more favorable conditions for the cooperative research work to be carried out. In June 1996, when Sun G. was invited by Mosbrugger to visit Tuebingen, Germany, for the 100th Anniversary of the Dept. Geology of the UT, Sun G. discussed again the Sino-German cooperation with Mosbrugger and introduced the story of his successful pursuit of "Prof. Obruchev's Road" in the Junggar Basin of Xinjiang, China. After the consultation and support by their close friend Dr. Ashraf A. R., the two sides agreed to undertake a cooperative study of the Mesozoic of Xinjiang, mainly focusing on the biostratigraphy and the ecosystems, as a new subject of Sino-German cooperation. Fortunately, at that time, the program for cooperation between the CAS and the Max-Planck Society was actively carried out. The Sino-German cooperative project on the study of Mesozoic stratigraphy and paleontology in Xinjiang, led by Sun G. and Mosbrugger, was successfully approved. Thus, the imaginations of Bonn and Tuebingen became reality in 1997 (Fig. 7).

## 1.2.2 Sino-German Cooperative Research Team in Xinjiang

Between 1997 and 2001, the cooperation research work led by Sun G. and Mosbrugger was generously supported by the Program of CAS (China) and the Max-Planck Society (Germany), and the NSFC (China). Since 2001, the cooperation has also been supported by the Sino-German Science Promoting Center (SGSPC). Moreover, with the support of the Xinjiang Bureau of Geology and Mineral Resources (XBGMR), the Sino-German Co-Working Station of Geosciences in Xinjiang (SGWSGX) was established in 2000, with the permanent office located in the Regional Geological Survey No. 1 of Xinjiang (RGSX-No. 1). Professors Zuo X. Y. (Ex-Director of the RGSX-No. 1), Sun G. and Mosbrugger V. were the co-directors of the Station. The SGWSGX work was kindly supported by the XBGMR in Urumqi. The establishment of the SGWSGX not only put forward the scientific research of geology and paleontology in Xinjiang but also greatly promoted the level of regional geological mapping and the international cooperation for the RGSX No. 1. During the SGWSGX working period, the staff of the RGSX No. 1 visited Germany for several times with young personnel for advanced study; and the German colleagues and students visited Xinjiang of China also for many times for the research cooperative work and geological excursions in Xinjiang. Thus, the SGWSGX has become a "service reception station" for all the members of the Station, and the Sino-German geoscience cooperation in Xinjiang rapidly reached a climax (Fig. 8).

In 2005, with the support of the SGSPC, the first "Sino-German Joint Laboratory for Paleontology and Geosciences" was established at Jilin University, China, which consisted of 13 scientific institutions from China and Germany, and Sun G. and Mosbrugger were appointed as co-Directors of the Laboratory. In October of 2005, the Opening Ceremony of the Laboratory was held in the RCPS of Jilin University in Changchun, and Prof. Han J. G. (Ex-Director of SGSPC on the Chinese side) and Academician Prof. Zhou Q. F. (Ex-President of Jilin Univ.) were invited to attend the ceremony. The new Sino-German laboratory still focused on the research cooperation of the Mesozoic of Xinjiang, China.

Over the years, with the development of the Sino-German cooperative research, several international symposia on the Mesozoic geology and paleontology of Xinjiang were held in Urumqi of Xinjiang, China, and the Sino-German cooperation has also expanded to include institutions of other countries concerned such as the United States and Japan. At the same time, several young scientists (including graduates) from China and Germany have been trained in the cooperative research in Xinjiang. In 2001, the SGWSGX station was identified as an excellent project of the CAS (China)-MPS (Germany). In 2010, Sun G. and Mosbrugger were honorably invited by the SGSPC to make a special report on the Sino-German cooperative research achievements at the 10th Anniversary of the SGSPC held in Beijing, which was the only research report in the celebration of the Anniversary (Fig. 62, B).

Chapter **2**

# Mesozoic strata of Xinjiang in general view

## 2.1 Overview of the three basins of Xinjiang

Mesozoic strata are widely distributed in Xinjiang, mainly concentrated in the Junggar-, Tarim- and Turpan-Hami (Tu-Ha) basins. The Tarim Basin covers an area of ca. 530, 000 km$^2$, ranking first; the Junggar Basin covers ca. 380, 000 km$^2$, ranking second; and the smaller is the Tu-Ha Basin, with an area of ca. 48, 000 km$^2$. The Tianshan Mountains lie in the center of Xinjiang, separating the Junggar Basin in the north from the Tarim Basin in the south, where both yield rich oil resources. The Tu-Ha Basin is located in the eastern Tianshan Mountains, extending from east to west, and has the largest coal reserves (e. g. the Shaerhu coalfield) known in China.

The three Mesozoic basins of Xinjiang were all formed on the Late Paleozoic folded tectogene, with mostly Upper Paleozoic basements. Several good sections show the Permian-Triassic (P-T) boundary (e. g. the Dalongkou of Jimsar in Junggar Basin), favorable for the study of the P-T boundary. The three basins of Xinjiang vary in structural type and their subsidence centers were constantly moving. For example, in the Junggar Basin from the Triassic to Jurassic and Cretaceous, the subsidence center was moving from north of Urumqi to the Changji (Manas River), and then to the central part of the basin. Tectonically, the intensities of movement in the three basins during the Mesozoic were also different. For example, in the Tarim Basin it was stronger in the west, while in the Junggar Basin it was stronger in the south, and in the Tu-Ha Basin it was stronger in the north (Kang, 2003). As the Mesozoic Xinjiang was covered with rivers, lakes, and swamps, there was an abundance of animals and plants there, which formed rich oil, coal, and natural gas deposits in the three basins, providing a valuable wealth for the reserve and development of fossil energy in China.

Because the geological settings, biological characters and evolution in the three Mesozoic basins in Xinjiang are basically consistent, their bio-strata can be correlated in general. Since the Junggar Basin is the most representative of the three basins, the authors take the strata of the Junggar Basin as the main criterion in this book, as seen in the unified Mesozoic stratigraphic table (Table 1).

## 2.2 Mesozoic strata in Xinjiang

The Mesozoic non-marine strata in Xinjiang are quite distinct: ① The Lower Triassic is dominated by red or variegated clastic deposits, the Middle Triassic is mainly composed of yellow-green fine clastic rocks with lacustrine deposits, and the Upper Triassic is dominated by gray and yellow gray fluvial-lacustrine deposits; ② The Lower and Middle Jurassic are dominated by gray coal-bearing deposits with a large number of commercially mined coal seams, while the Upper Jurassic is dominated by variegated or purple deposits with abundant dinosaur fossils; and ③ In the Cretaceous, the Lower Cretaceous is basically composed of grayish yellow and grayish green deposits and variegated deposits appeared alternately, with missing of the lowest Cretaceous in deposition; while the Upper Cretaceous mainly consists of purple

Tabel 1 Simplified chart of Mesozoic strata of Xinjiang sampled with the Junggar Basin

| | | | Stratiphic units | Main rocks | Note |
|---|---|---|---|---|---|
| Cretaceous | K<sub>2</sub> | | Lower Ziniquanzi Fm | grayish green and red brown sandstone, basally white sandstone and conglomerate | fossil ostrocods, etc. |
| | | | Donggou Fm | mainly purple sandstone and conglomerate | dinosaurs, etc. |
| | K<sub>1</sub> | Tugulu Gr. | Lianmuqin Fm | mainly variegated sandstone, mudstone | fossil bivalves, etc. |
| | | | Shengjinkou Fm | grayish green-yellow sandstone, mudstone | fossil fish, etc. |
| | | | Hutubi Fm | mainly purple sandstone, mudstone | fossil bivalves, etc. |
| | | | Qingshuihe Fm | grayish green-yellow sandstone, mudstone | pterosauirs, etc. |
| Jurassic | J<sub>3</sub> | | Kalaza Fm | mainly purple red conglomerates | *Hudiesaurus* in Tu-Ha Basin |
| | | | Qigu Fm | mainly purple and variegated sandstone | *Xinjiangotitan*, etc. |
| | J<sub>2</sub> | | Toutunhe Fm | mainly yellow-green sandstone, siltstone | dinosaur tracks |
| | | | Xishanyao Fm | mainly gray sandstone and siltstone | plant fossils, coal |
| | J<sub>1</sub> | | Sangonghe Fm | mainly yellow sandstone, conglomerate | plant fossils |
| | | | Badaowan Fm | mainly gray sandstone and siltstone, basally grayish white conglomerate | plant fossils, coal |
| Triassic | T<sub>3</sub> | | Haojiagou Fm | mainly gray sandstone, siltstone | plant and limulus fossils, coal in upper |
| | | | Huangshanjie Fm | mainly grayish yellow sandstone and conglomerate | plant fossils |
| | T<sub>2</sub> | | Karamay Fm | mainly grayish green fine sandstone and siltstone | fossil fish, etc. |
| | T<sub>1</sub> | Upper Cangfanggou Gr. | Shaofanggou Fm | mainly variegated sandstone and conglomerate | sporpollen, ostrocods, etc. |
| | | | Jiucaiyuan Fm | mainly purple red sandstone and conglomerate | *Lystrosaurus*, etc. |

or yellowish gray deposits, usually with rich dinosaur and other fossils, except for the southwestern edge of the Tarim Basin where some marine deposits still remained (Xi et al., 2019) (Table 1).

## 2.2.1 Triassic

### (1) Lower Triassic

The Lower Triassic non-marine strata in Xinjiang are represented mainly by red or variegated clastic sediments, yielding vertebrate fossils such as *Lystrosaurus*. The strata are conformably overlying the Upper Permian, and a spectacular P/T section is exposed on the east bank of the river in Dalongkou village, about 15 km south of Jimusar Town in southeastern Junggar Basin (Figs. 12; 13; 66).

According to the studies of the XBGMR (1993) and Li et al. (2003), the stratigraphic sequence of the P/T boundary section in the Dalongkou is as follows:

thickness (unit: m)

5. Shaofanggou Formation ($T_{1s}$) : purple sandstone and siltstone,
   yielding sporopollen ------------------------------------------260–311
4. Jiucaiyuan Formation ($T_{1j}$) : purple silty mudstone intercalated with
   yellow-green siltstone, yielding sporopollen and the therapsid *Lystrosaurus* ------220–282
3. Guodikeng Formation (P-T) : grayish green and purple silty mudstones
   intercalated with grayish black and variegated siltstone and mudstone,
   yielding sporopollen, bivalves, ostracods and conchostracans---------------- 144
2. Wutonggou Formation ($P_{3w}$) : gray and grayish black sandstone, siltstone
   and mudstone, yielding sporopollen and other fossils--------------------- 220
1. Quanzijie Formation ($P_{3q}$) : purple and grayish green siltstone,
   mudstone intercalated with conglomerates, yielding *Callipteris-Comia-*
   *Inopteris* plant assemblage and bivalves ------------------------------122–242

The Guodikeng Formation is the P-T boundary strata containing the PTB. Overlying is the Jiucaiyuan Formation ($T_{1j}$), and underlying is the Wutonggou Formation ($P_{3w}$) which underlay the Quanzijie Formation ($P_{3q}$) yielding the Late Permian Angaran flora represented by the assemblage of *Callipteris-Comia-Inopteris*, as well as bivalves and vertebrate fossils. Certainly, the PTB at Dalongkou deserves further study.

## (2) Middle Triassic

The Middle Triassic of the Junggar Basin is represented by the Karamay Formation, composed mainly of grayish green and yellow sandstone, intercalated with mudstone and conglomerate, with an about 3.8 m thick fine conglomerate at the base of the formation. The total thickness of the formation is between 265-885 m. The stratotype section of the Karamay Formation is located near Karamay City in the northwestern Junggar Basin, and a typical section is exposed in the Xiaoquangou of Fukang County, southeastern Junggar Basin, showing the Karamay Formation overlying the Shaofanggou Formation ($T_{1s}$), and underlying the Huangshanjie Formation ($T_{3hs}$), both in conformity. The Karamay Formation is rich in fossils of plants, bivalves, conchostracans, fish, and tetrapods, such as the plants *Neocalamites carcinoides, N. carrerei, Bernoullia zeilleri, Daenaeopsis fecunda, Chiropteris?*

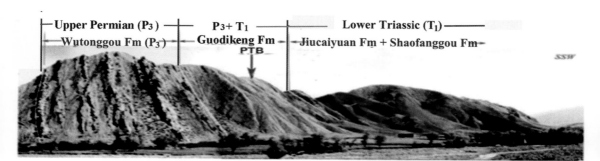

Fig. 66  Dalongkou section with P-T boundary (after Ashraf et al., 2004 )

*yuani*, *Glossophyllum shensiense*, *Thinnfeldia nordenskioldi*; bivalves *Ferganoconcha sibirica*, *F. rotunda*, *Sibiriconcha anodontites*; conchostracans *Mesolimnadiopsis karamaica*, *Lioestheria euenkiensis*; fish *Sinosemionodus urumuchi*; and tetrapods *Parakannemeyeria brevirostris*, *Parotosaurus* sp. (XBGMR, 1993) (Fig. 14).

## (3) Upper Triassic

The Upper Triassic in Xinjiang comprises mainly fluvial or fluvio-lacustrine deposits, including the Huangshanjie Formation (T$_{3hs}$) and the Haojiagou Formation (T$_{3hj}$) in ascending order. A typical section of the Huangshanjie Formation is located in Xiaoquangou of Fukang, composed mainly of gray and grayish-green mudstone intercalated with siltstone, about 153 m thick and rich in fish and other fossils, such as *Bogdania fragmenia*, *Fukangichthys longidorsa*, *Fukangolepis barbavos* (XBGMR, 1993)

The best and stratotype section of the Haojiagou Formation is located in Haojiagou about 50 km west of Urumqi. The Haojiagou Formation (named by XBGMR in 1959 ) is composed of gray sandstone, siltstone, and a small amount of conglomerate, with multi-layer carbonaceous siltstone and coal seams in its upper part, representing lacustrine and partially fluvial conditions, and exposed in the beds 20-41 of the section (Fig. 67). The formation is characterized by rich fossil plants represented by the assemblage of *Glossophyllum-Cycadocarpidium* (Sun et al., 2001b, 2010). The formation is underlying the Lower Jurassic Badaowan Formation (J$_{1b}$) in conformity as evident from the grayish black siltstone at the top of the Haojiagou Formation and the gray to white glutenite at the base of the Badaowan Formation (bed 42), which provide favorable conditions for studying the T-J boundary (Figs.15; 67). In 1997, the authors measured the Haojiagou Formation to be 280 m in thickness in this section (Figs. 15).

The Huangshanjie Formation in this section is mainly composed of yellowish gray sandstone, siltstone, and a small amount of conglomerate, showing fluvial facies, underlying in conformity the Haojiaogou Formation and yielding rich plant fossils highlighted by *Danaeopsis-Nanzhangophyllum* assemblage (beds 1-19 in Fig. 67, A).

**Badaowan Fm**

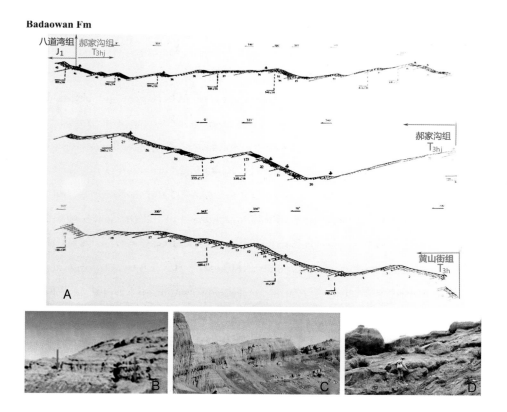

Fig. 67  Upper Triassic strata from Haojiagou geological section

A. The Haojiagou section measured by the authors: beds 1-19 showing the Huangshanjie Formation, beds 20-41 showing the Haojiagou Formation (after Sun et al., 2001b); B. The lower part of the Huangshanjie Formation ($T_{3hs}$); C. The upper part of the Haojiagou Formation ($T_{3hj}$); D. The top part of the Haojiagou Formation and its overlying base of the Lower Jurassic Badaowan Formation ($J_{1b}$, bed. 42).

## 2.2.2 Jurassic

The Jurassic strata in Xinjiang are mainly fluvial and lacustrine deposits: the Lower and Middle Jurassic are mainly composed of gray sandstone and siltstone intercalated with gray to black thick coal-bearing deposits. The Upper Jurassic is mainly composed of purplish red sandstone, siltstone, and variegated sandstone. According to the data of the XBGMR (1993) and the authors' work, the detailed composition of the Jurassic is as the follows (taking the Junggar Basin as an example).

### (1) Lower Jurassic

In the Junggar Basin (in ascending order), the lowest unit is Badaowan Formation ($J_{1b}$) which is coal-bearing, and mainly composed of gray sandstone, with grayish white conglomerate and sandstone at the base of the formation, showing a fluvial and swamp facies, with a thickness of about 622 m. The Badaowan Formation is rich in fossils, represented by the plants *Dictyophyllum* and *Coniopteris*, and the bivalves *Unio* and *Ferganoconcha*. The typical section of this formation is in the Badaowan Coal Mine east of Urumqi, and it is also well exposed in the Haojiagou section in which the base of the Badaowan Formation is characterized by gray white conglomerate and sandstone overlying the Upper Triassic Haojiagou Formation ( Sun et al., 2001b, 2010) (Fig. 16, B, C; Fig. 67, A in Bed 42).

The second deposit of the Lower Jurassic in the Junggar Basin is the Sangonghe Formation ($J_{1s}$) , composed mainly of grayish yellow and variegated sandstone in lacustrine facies, with thick grayish yellow conglomerate layers in some areas. The formation has a thickness of 882 m, yielding fossils such as plants *Equisetites laterale* (Phill.) Morris, *Coniopteris hymenophylloides* Brongniart, etc. (Sun et al., 2001b) ; the bivalves *Ferganoconcha*, *Sibiriconcha*, and *Unio* (XBGMR, 1993), and the conchostracans *Palaeolimnadia baitianbaensis* Chen, and *Euestheria tenuiformis* (Zaspelova), etc. (Chen et al., 1998) (Fig. 16, E).

## (2) Middle Jurassic

The Middle Jurassic strata are the main coal-bearing strata in Xinjiang, represented by the Xishanyao Formation ($J_{2s}$) in the Junggar Basin. The Xishanyao Formation is mainly composed of gray, grayish to yellow sandstone and siltstone, with thick coal-bearing sections, and has a total thickness of about 980 m, mainly showing lacustrine swamp facies. The stratotype section of this formation is located in the Xishanyao Coal Mine in the western suburbs of Urumqi. The typical sections of the formation are exposed in the Baiyang River area of Tuoli in the northwestern Junggar Basin and in the Manas River area west of Urumqi. In the Tu-Ha Basin, typical sections of the Middle Jurassic are found in the Shaerhu Coalfield, one of the largest coal mines in China. The Xishanyao Formation yields abundant fossil plants represented by the famous Middle Jurassic Baiyanghe flora and Shaerhu flora (Figs. 31-34; 43-48; 68, A-C).

The Toutunhe Formation ($J_{2t}$) is overlying the Xishanyao Formation in conformity in the Junggar Basin. The formation is mainly composed of grayish green, yellow and purple siltstone and mudstone, about 654 m thick, showing lacustrine facies. The formation yields abundant fossils, including ostracods, bivalves, conchostracans, plants, etc. Its stratotype section is located in the Toutunhe, west of Urumqi. The lower part of the Shishugou Formation in the Jiangjunmiao area of Qitai may be coeval with the Toutunhe Formation, yielding rich fossil wood. The Sanjianfang Formation in the Tu-Ha Basin may also be coeval with the Toutunhe Formation, yielding the largest known Jurassic dinosaur tracks in the Shanshan Gobi (Figs. 37; 68, D-G). According to the isotopic study by Deng et al. (2015), the Toutunhe Formation is dated between 166.2 Ma and 160.8 Ma, and the main depositing time is the late Middle Jurassic (Deng et al., 2015).

## (3) Upper Jurassic

The Upper Jurassic in Xinjiang is represented by (in ascending order) the Qigu

Formation (J$_{3q}$) and the Kalaza Formation (J$_{3k}$).

Qigu Formation is mainly composed of purplish red and green sandstone and mudstone, about 683 m in total thickness. The stratotype section of this formation is located in Qigu of the Hutubi County, northwestern Junggar Basin. The formation is characterized by abundant dinosaurs and other vertebrate fossils, such as the giant dinosaur *Mamenchisaurus sinocanadorum* from Jiangjunmiao of Qitai, and the early

Fig. 68 Middle Jurassic strata in Xinjiang

A-C. Middle Jurassic Xishanyao Formation in Baiyanghe (Junggar Basin); B, C. Middle Jurassic Xishanyao Formation in Shaerhu Coal-Mine (Turpan-Hami Basin), E-G. Upper part of Middle Jurassic strata in Xinjiang: D showing the Shishugou Formation in Jiangjunmiao of Qitai (Junggar Basin); E and F showing the Sanjianfang Formation in Shanshan (Turpan-Hami Basin); G showing the Toutunhe Formation (J$_{2t}$) in Junggar Basin

mammal *Dsungarodon* from Liuhuanggou of Urumqi, which are both in the Junggar Basin. In the Tu-Ha Basin, the Qigu Formation yielded the largest known Jurassic dinosaur *Xinjiangotitan shanshanensis* and turtle *Xinjiangchytes wusu* from Shanshan, both found by the authors (Figs. 18, 69).

Regarding the age of the Qigu Formation, Wang and Gao (2012) reported a SHRIMP U-Pb dating of zircons 164.6 ± 1.4 Ma from the Qigu Formation in Honggou valley of Hewan County, south of Shihezi and west of Uqumqi in Junggar Basin (Wang

Fig. 69  Upper Jurassic strata in Xinjiang

A-D. Upper Jurassic Qigu Formation in Jiangjunmiao of Qitai, Junggar Basin, B and C showing the scenes of investigation of dinosaur fossils by the Sino-German research team (2004); E-I. Upper Jurassic Qigu Formation in Qiketai of Shanshan, Tu-Ha Basin, and showing the Sino-German research team excavating dinosaur fossils *Xinjiangotitan* (2011-2012)

and Gao, 2012). After that, Deng et al. (2015) reported dating of the Qigu Formation as ca. 155.3 Ma representing the top deposit age of the formation, and suggested the depositing duration as about 5.5 Ma for this formation, covering the upper Oxfordian and low-mid Kimmeridgian (Deng et al., 2015).

The Kalaza Formation ($J_{3k}$) is conformably overlying the Qigu Formation, and is mainly composed of grayish brown conglomerate and sandstone, with between 50-80 m in thick, mainly in fluvial facies (Fig. 19). The stratotype section of this formation is located in the Kalaza mountains, Changji City near Urumqi. Some sporopollen fossils were reported from this formation. In the Junggar Basin, the Kalaza Formation is overlain by the Lower Cretaceous Qingshuihe Formation of the Tugulu Group in unconformity (BGMRX, 1993; Zhang et al., 2021). Deng et al. considered the duration of the unconformity lasted over 7 Ma (Deng et al., 2015).

### 2.2.3 Cretaceous

In general, the study of the Cretaceous strata of Xinjiang has not been comprehensive so far.

### (1) Lower Cretaceous

In the Junggar Basin, the Lower Cretaceous Tugulu (Tugulik) Group is mainly composed of gray, green, purple, and other variegated sediments, up to 927.5 m thick. The Group is divided into four formations (in ascending order): ① Lianmuqin Formation: mainly variegated sandstone and mudston; ② Shengjinkou Formation: mainly grayish green and grayish yellow sandstone and mudstone; ③ Hutubi Formation: mainly purple sandstone and mudstone; and ④ Qingshuihe Formation: grayish green, yellowish green sandstone and mudstone with a basal conglomerate. The Tugulu Group is characterized by lacustrine deposits yielding ostracods, fish, dinosaurs, and pterosaurs. The well-known pterosaur *Dsungaripterus weii*, the first complete pterosaur fossil found in China, was studied first by Young (1964), and newly studied by Wang et al. who have found many new pterosaurs, particularly the new taxon *Hamipterus* in Hami of the Tu-Ha Basin, also (Wang et al., 2014, 2017) (Figs. 20, A-C; 70, A). The age of the Tugulu Group has been considered Aptian-Albian (BGMRX, 1993; Wu and Liu, 1988).

However, in recent years, according to the study of charophytes and other fossils, the age of the Tugulu Group in the Junggar Basin has been considered Berriasian-Barremian (Yang et al., 2008; Xi et al., 2019). In the eastern Tarim Basin, the Lower Cretaceous Kapushaliang Group and the Bashijiqike Formation are considered as the early-middle and late Early Cretaceous respectively (Xi et al., 2019).

## (2) Upper Cretaceous

The Upper Cretaceous in the Junggar Basin is represented by the Donggou Formation, which is mainly composed of grayish purple and variegated sediments with a thickness of up to 884.5 m (Figs. 20, H; 70, B). This formation yields abundant dinosaurs and other vertebrate fossils found in the Wurhe (the "Ghost city") -Sangequan (the "Three springs") area in the northern Junggar Basin. According to the fossil ostracods, the Donggou Formation is probably Coniacian-Campanian in age (Zheng et al., 2013), while some dinosaur egg fossils (e.g. *Ooelithes elongatus*) from the upper part of the Donggou Formation in west Queergou of Hutubi, suggest Late Cretaceous age (Zhao, 1980). As for the Ziniquanzi Formation, its age has been controversial and the ostracods and charophytes suggested that its lower part belongs to the uppermost Cretaceous, while the upper part represents the Cretaceous-Paleocene transition, although the K-Pg boundary strata are missing (Yang and Shen, 2004; Zheng et al., 2013; Xi et al., 2019).

In Tu-Ha Basin, the Upper Cretaceous yielded also dinosaurs, such as *Shanshanosaurus* (Dong, 1993).

During the Middle-Late Cretaceous, the western part of the Tarim Basin was a horn-shaped bay opening westward, representing a branch of the eastern *paleo*-Tethys

Ocean. The marine or intercalated marine/non-marine strata consist of (in ascending order) the Lower Cretaceous Kizilsu Group (Barremian-Albian), the Upper Cretaceous Kukebai Formation, Igeziya Formation (Cenomanian-Maastrichtian), and the Tuylock Formation (the K-Pg interval strata). All the above-mentioned strata yield abundant fossils such as foraminifera, ostracods, dinoflagellates, sporopollen, bivalves and gastropods, and a small number of ammonites, brachiopods, sea urchins, and shark teeth. The transgression in the Tarim Basin began in the late Aptian to early Albian, and represents the larger scaled transgression since the Cenomanian. There were five large-scale transgressions and regressions in this area during the Late Cretaceous-Paleocene. Until ca. 41 Ma, the sea withdrew from the Kunlun Mountains, and at ca. 34 Ma withdrew from the Tianshan Mountains (Xi et al., 2020) (Fig. 21).

Fig. 70 Cretaceous strata in Xinjiang

A. Lower Cretaceous Tugulu Group in Hami of Tu-Ha Basin (after Wang et al., 2014); B. Upper Cretaceous Donggou Formation yielding dinosaurs in Wurhe of Junggar Basin

# Chapter 3

## New discoveries of Mesozoic fossils in Xinjiang

Since 1997, the Sino-German research team has carried out cooperative study of the Mesozoic in Xinjiang for more than 20 years, starting from the Junggar Basin, then turning to the Tu-Ha Basin, and partly the Tarim Basin, and making a series of new discoveries in paleontology and stratigraphy. The new discoveries of Mesozoic fossils include dicynodontid therapsid, dinosaurs, pterosaurs, turtles, crocodiles, sharks, mammals, and plants (Martin et al., 2010b). The main achievements are represented by the first discoveries of Late Triassic Haojiagou flora, the largest Jurassic dinosaur *Xinjiangotitan* and dinosaur track group, Late Jurassic new mammal assemblage, the new species of Jurassic turtles *Xinjiangchyles*, the Early Triassic synapsidan *Lysrosaurus*, and the new round of research on the Middle Jurassic Baiyanghe flora and Shahu flora (Fig. 71).

# 3.1 Early Triassic *Lystrosaurus* from Dalongkou

## 3.1.1 The P-T boundary in Dalongkou

The Permian-Triassic (P-T) boundary is a well-known stratigraphic line world-wide, showing the mass extinction and recovery at ca. 252 Ma. At that time interval, the largest mass extinction event occurred on earth since the beginning of the Phanerozoic, which accounted for 85% of the total taxa, and new life began to evolve since the Early Triassic (the beginning of Mesozoic). Therefore, the study of the fossils and strata of this period is of great significance for understanding the changes of the earth's biology and environment.

Fig. 71 Sketch map showing the main localities of Mesozoic fossils newly found in Xinjiang

In the southeastern Junggar Basin, the P-T boundary and its related strata are spectacularly exposed at the Dalongko village of Jimsar near Fukang City (Fig. 13). Under the boundary strata (i. e. Guodikeng Formation, P-T), the Wutonggou Formation ($P_{3w}$) yields sporopollen fossils indicating a Late Permian age, and further below the Quanzijie Formation ($P_{3q}$) yields the Late Permian Angaran flora represented by the fossil plant assemblage of *Callipteris-Comia-Inopteris*, and bivalves and vertebrates. Above the boundary strata (Guodikeng Formation), the overlying Jiucaiyuan Formation ($T_{1j}$) yields the index vertebrate fossil *Lystrosaurus* characteristic of the Early Triassic. Therefore, the Guodikeng Formation (P-T) has become a "hot spot" for geoscientists world-wide. In the 1980s, a China-US cooperative research team guided by the Chinese Academy of Geosciences (CAGS) carried out cooperative research work on the P-T boundary here, but the work was interrupted due to some disagreement. Thus, the arrival of the Sino-German research team has breathed new vitality into the research on the P-T boundary in the Dalongkou section.

## 3.1.2 Discovery of *Lystrosaurus*

*Lystrosaurus* is a dicynodontid therapsid (Synapsida), with a big head, short neck and barrel-shaped body, and has the size of a pig. Its snout is bent down, the skull and orbit are higher, and the body structure has some progressive characters seen in mammals, but the head still remains primitive. Except for two enlarged upper canines, the jaws are toothless and possess a horny beak in the living animal. According to the analysis of its shoulder and pelvic girdles and limbs, *Lystrosaurus* has a sprawling gait. Its special skull shape is seen as an adaptation for ingesting hard plants as they grow in arid environments. The strong forefeet are suitable for digging. The bone histology suggests that some species may have been semi-aquatic. *Lystrosaurus* survived the P-T boundary and is one of the index fossils for the Early Triassic, and has been found in South Africa, India, Russia, and Xinjiang of China. An almost complete skeleton of a juvenile *Lystrosaurus* was discovered in 2002 in the lowermost Jiucaiyuan Formation (Lower Triassic) at Dalongkou by Maisch M. and Matzke A., the two young experts of the authors' research team (Fig. 22). The material is one of the most complete *Lystrosaurus* skeletons known and probably the most complete from China. It certainly is one of the best juvenile dicynodont skeletons found outside Africa (Maisch et al., 2004). The juvenile age of the skeleton from Dalongkou is indicated by its small size, minor ornamentation of the dorsal skull roof, and non-classified elements of the shoulder and pelvic girdles. The neural arches are not fused with the centra of the vertebrae and carpus and tarsus are very weakly ossified. Besides the juvenile skeleton, several skulls and many postcranial skeletal elements of *Lystrosaurus* have also been unearthed by the team.

# 3.2  Late Triassic Haojiagou flora

The Late Triassic Haojiagou flora was found by the authors between 1997 and 2001 from the stratotype section of the Upper Triassic strata in the Haojiagou valley about 50 km west of Urumqi (Sun et al., 2001b, 2010) (Figs. 15; 67). The flora consists of two plant assemblages, including the Huangshanjie Assemblage of *Danaeopsis-Nanzhangophyllum*, in the Carnian-Norian age (Fig. 23), and the Haojiagou Assemblage of *Glossophyllum-Cycadocarpidium*, in Norian-Rhaetian age (Fig. 24). The representative composition of the Haojiagou flora includes:

### Equisetales

1. *Neocalamites* sp.

### Filicinae

2. *Todites?* sp.

3. *Danaeopsis tenuinervis* Sun, Mosbrugger et Li

4. *Danaeopsis* sp.

5. *Cladophlebis grabauiana* Sze

6. *Cladophlebis kaoiana* Sze

7. *Cladophlebis* sp.

**Pterispermae**

8. *Thinfeldia? minisecta* Sun, Mosbrugger et Li

9. *Thinnfeldia* sp.

10. *Nanzhangophyllum haojiagouense* Sun, Mosbrugger et Li

**Cycadophytes ?**

11. *Taniopteris mucronulata* Sun, Mosbrugger et Li

12. *Taniopteris* sp.

**Ginkgoales**

13. *Ginkgoidium sp.*

14. *Sphenobaiera? sp.*

15. *Glossophyllum shensiense* Sze

16. *Solenites haojiagouensis* Yang et al.

**Coniferales**

17. *Cycadocarpidium erdmanni* Nathorst

18. *Cycadocarpidium swabii* Nathorst

19. *Cycadocarpidium* sp.

## 3.2.1  Late Triassic Huangshanjie Assemblage

The Huangshanjie Assemblage (*Danaeopsis-Nanzhangophyllum* Ass.) was found in the Huangshanjie Formation consisting of grayish yellow and yellowish grayish siltstone in the lower part of the formation. The main taxa include: *Neocalamites* sp., *Danaeopsis tenuinervis* Sun, Mosbrugger et Li, *Danaeopsis* sp., *Cladophlebis grabauiana* Sze, *C. kaoiana* Sze, *Thinfeldia? minisecta* Sun, Mosbrugger et Li, *Nanzhangophyllum haojiagouense* Sun, Mosbrugger et Li, *Taniopteris mucronulata* Sun, Mosbrugger et Li, and *Taniopteris* sp. (Figs. 23; 72). Among them, *Danaeopsis* is a common genus of the Late Triassic flora in China and a most representative taxon of the Late Triassic flora in Northern China. *Cladophlebis grabauiana* and *C. kaoiana* are important taxa of the Late Triassic Yenchang flora in Northern China. *Nanzhangophyllum* is a common member of the Late Triassic flora in Southern China, mainly found in the Upper Triassic of Nanzhang, Hubei Province. Therefore, the Huangshanjie Assemblage exhibits clearly the features of a Late Triassic flora (Sun et al., 2001b, 2010).

## 3.2.2 Late Triassic Haojiagou Assemblage

The plants of the Haojiagou Assemblage (*Glossophyllum-Cycadocarpidium* Ass.) were found in light gray and gray siltstone intercalated with coal seams in the Upper Triassic Haojiagou Formation. The main components of the assemblage include *Todites?* sp., *Cladophlebis* sp., *Thinnfeldia* sp., *Ginkgoidium* sp., *Sphenobaiera?* sp., *Glossophyllum shensiense* Sze, *Solenites haojiagouensis* Yang et al., *Cycadocarpidium erdmanni* Nathorst, *C. swabii* Nathorst, and *Cycadocarpidium* sp. Among them, *Glossophylum shensense* is the most representative taxon of the Late Triassic flora in Northern China. *Cycadocarpidium erdmani* and *C. swabii* are both index fossils for the Late Triassic flora in the world, mainly indicating Norian-Rhaetian age. Therefore, the Haojiagou assemblage shows Late Triassic age, and is slightly younger than the Huangshanjie assemblage (Sun et al., 2001b, 2010) (Figs. 24; 72).

In floral characters, the Late Triassic Haojiagou flora is more similar to the Yenchang flora of northern Shaanxi, China, generally, based on the floral composition including many components of the Yenchang flora, such as *Danaeopsis* sp., *Glossophyllum shensiense*, *Cladophlebis grabauiana*, *C. kaoiana*, and *Neocalamites* sp., which belongs to the Late Triassic *Danaeopsis-Glossophyllum* flora of Northern China. However, since the Haojiagou flora has some common elements with the Late Triassic flora of Southern China, such as *Nanzhangophyllum*, *Cycadocarpidium erdmanni, C. swabii,* and a lot of *Taniopteris,* the Haojiagou flora has a somewhat mixed aspect of the Late Triassic Northern and Southern floras of China. The occurrence of the mixed characters may be a result of the paleogeographic position of

Fig. 72  Late Triassic Haojiagou flora from Haojiagou section in Junggar Basin

A-M. Late Triassic Haojiagou flora: A. *Cladophlebis kaoiana*; B. *C. grabauiana*; C. *Danaeopsis tenuinervis*; D, E. *Nanzhangophyllum haojiagouense*; F. *Taniopteris mucronulata*; G. *Thinfeldia? minisecta*; H. *Ginkgoidium* sp.; I, J. *Glossophyllum shensiense*; K. *Cycadocarpidium swabii*; L. *C. erdmanni*; M. *Cycadocarpidium* sp.; All bar =1 cm (A-M after Sun et al., 2001b, 2010); N. Upper Triassic Huangshanjie Formation; O, P. Upper Triassic Haojiagou Formation

the Haojiagou flora which was located in the southwestern Junggar Basin, not far from the paleophytogeographic border between the Southern and Northern Floras of China during the Late Triassic (Sun, 1987, 1993a; Sun et al., 1995). Thus, the discovery of the Late Triassic Haojiagou flora in Xinjiang is very significant for the further study of the distribution and evolution of the Late Triassic flora in Xinjiang and the Late Triassic phytogeography of China.

### 3.2.3 Palynoflora

From 1998 to 2000, Ashraf et al. of the authors' research team studied the Late Triassic palynoflora in the Haojiagou section in detail, and collected 217 samples of sporopollen of which 48 were from the Huangshanjie Formation and 169 from the Haojiagou Formation (Ashraf et al., 1998, 2001, 2010).

Palynomorphs of 93 species belonging to 57 genera in total were identified for the Huangshanjie and Haojiagou formations, including 58 species of 41 genera from the Huangshanjie Formation, and 52 species of 37 genera from the Haojiagou Formation. The taxa of the Late Triassic palynoflora are shown in Table 2 (Ashraf et al., 1998).

The Upper Triassic Huangshanjie Formation is characterized by the assemblage of *Concavisporites-Duplexisporites problematicus-Lophotrites sangburensis-Cyclotrites oligografinifer* zone, mainly showing the Norian age; while the Haojiagou Formation is characterized by the assemblage of the *Concavisporites-Duplexisporites problematicus-Ruccisporites tuberculatus* zone, mainly showing the Rhaetian age (Ashraf et al., 1998, 2001, 2004, 2010) (Figs. 25; 26).

Ecologically, the palynoflora shows the basic hygric types: ① Hygrophytic forms: these presumably include the Bryophytes, Selaginellaceae, Lycopodiaceae, and Osmundaceae, which prefer relatively moist (soil) conditions; ② Meso-hygrophytic forms: they essentially include the other ferns such as the Dicksoniaceae,

Table 2 Overview of the palynomorphic taxa in the Huangshanjie and Haojiagou Formations

(after Ashraf et al., 1998, 2001; with the authors' revisions)

| No. | Taxa | Huangshanjie Fm | Haojiagou Fm |
|-----|------|:---------------:|:------------:|
| | Bryophytes /Pteridophytes | | |
| 1 | *Stereisporites perforatus* | | X |
| 2 | *Lycopodiacidites* sp. | X | |
| 3 | *Aratrisporites fischeri* | | X |
| 4 | *Aratrisporites scabratus* | | X |
| 5 | *Aratrisporites strigosus* | | X |
| 6 | *Aratrisporites* sp. | | X |
| 7 | *Densoisporites velatus* | X | |
| 8 | *Osmundacidites wellmanii* | X | |
| 9 | *Osmundacidites wellmanii conatus* | X | |
| 10 | *Osmundacidites* sp. | | X |
| 11 | *Todisporites concentricus* | X | |
| 12 | *Auritulinasporites intrastriatus* | X | |
| 13 | *Concavisporites bohemiensis* | X | |
| 14 | *Concavisporites mortoni* | | X |
| 15 | *Concavisporites toralis* | | X |
| 16 | *Concavisporites* sp. | X | |
| 17 | *Dictyophyllidites mortoni* | X | X |
| 18 | *Dictyophyllidites* sp. | | |
| 19 | *Annulispora folliculusa* | X | |
| 20 | *Apiculatisporis* sp. | X | X |
| 21 | *Baculatisporites comanmensis* | X | |
| 22 | *Carnisporites granulatus* | X | |
| 23 | *Camarozonosporites rudis* | X | |
| 24 | *Cyclogranisporites* sp. | X | |
| 25 | *Duplexisporites gyrates* | X | |
| 26 | *Duplexisporites problematicus* | X | |

(Table 2 continued)

| No. | Taxa | Huangshanjie Fm | Haojiagou Fm |
|-----|------|-----------------|--------------|
| 27 | *Duplexisporites scanicus* | | X |
| 28 | *Duplexisporites* sp. | X | |
| 29 | *Habrozonosporites decorates* | X | |
| 30 | *Habrozonosporites* sp. | X | |
| 31 | *Hsuisporites liaoningensis* | X | |
| 32 | *Kraeuselisporites reissingeri* | X | |
| 33 | *Limatulasporites haojiagouensis* | | X |
| 34 | *Limatulasporites* sp. | | X |
| 35 | *Polycingulatisporites simplex* | | X |
| 36 | *Polycingulatisporites* sp. | X | |
| 37 | *Punctatisporites* sp. | X | |
| 38 | *Reticulatisporites* sp. | X | |
| 39 | *Retusotriletes mesozoicus* | X | |
| 40 | *Retusotriletes* sp. | | X |
| 41 | *Verrucosisporites* sp. | X | |
| | Gymnospermae | | |
| 42 | *Abietineaepollenites* sp. | | X |
| 43 | *Alisporites australis* | X | |
| 44 | *Alisporites thomasii* | X | |
| 45 | *Alisporites* sp. | | X |
| 46 | *Chasmatosporites apertus* | | X |
| 47 | *Chasmatosporites anagrammensis* | | X |
| 48 | *Chasmatosporites hianu* | X | |
| 49 | *Chasmatosporites verruculosus* | X | |
| 50 | *Chasmatosporites* sp. | | |
| 51 | *Ginkgocycadophytus nitidus* | X | |
| 52 | *Cycadopites fragilis* | X | |
| 53 | *Cycadopites dilucidus* | X | |

(Table 2 continued)

| No. | Taxa | Huangshanjie Fm | Haojiagou Fm |
|---|---|---|---|
| 54 | *Cycadopites reticulatus* | X | |
| 55 | *Cycadopites rugugranulatus* | X | |
| 56 | *Cycadopites subgranulosus* | X | |
| 57 | *Cycadopites tivoliensis* | X | X |
| 58 | *Cycadopites* sp. | | X |
| 59 | *Ovalipollis* sp. | X | |
| 60 | *Piceites expositus* | X | |
| 61 | *Piceites* sp. | X | |
| 62 | *Pinuspollenites* sp. | | |
| 63 | *Striatoabietites aytugii* | X | |
| 64 | *Platysaccus lopsinensis* | X | |
| 65 | *Platysaccus* sp. | X | |
| 66 | *Paleoconiferus asaccatus* | X | |
| 67 | *Podocarpites multisimus* | | X |
| 68 | *Podocarpites* sp. | | X |
| 69 | *Protopinus* sp. | X | |
| 70 | *Psophosphaera* | X | |
| 71 | *Cordaitina* sp. | X | |
| 72 | *Florinites* sp. | X | |
| 73 | *Pseudocrustaesporites* sp. | X | |
| 74 | *Tetrasaccus* sp. | X | |
| 75 | *Quadraeculina limbata* | | X |
| 76 | *Chordasporites singulichorda* | X | |
| 77 | *Chordasporites* sp. | | X |
| 78 | *Lunatisporites rhaeticus* | X | X |
| 79 | *Protohaploxypinus* sp. | X | |
| 80 | *Taeniaesporites combinatus* | X | X |

(Table 2 continued)

| No. | Taxa | Huangshanjie Fm | Haojiagou Fm |
|-----|------|-----------------|--------------|
| 81 | *Taeniaesporites xingxianensis* | X | |
| 82 | *Taeniaesporites leptocorpus* | | X |
| 83 | *Taeniaesporites* sp. | X | X |
| 84 | *Vittatina* sp. | X | |
| | Megaspores | | |
| 85 | *Echitriletes prerussus* | | X |
| 86 | *Otyniosporites* sp. | | X |
| 87 | *Aneuletes* sp. | | X |
| 88 | *Cabochonicus carbunculus* | | X |
| 89 | *Cabochonicus* sp. | | X |
| 90 | *Horstisporites* sp. | | X |
| 91 | *Bothriotriletes* sp. | | X |
| 92 | *Paxillitriletes* sp. | | X |
| 93 | *Calamospora rhaeticus* | | X |

Dipteridaceae/Matoniacae and Schizaeaceae, which characterize somewhat drier locations than the group of ① ; ③ Mesophytic forms: with the cycads, bennettitaleans, ginkgos' relatives and conifers, which represent the driest conditions in comparison.

This mixture of hygrophytic and meso-hygrophytic on the one hand, and mesophytic forms on the other hand, which is quite balanced in terms of diversity, can reflect either a mixture of corresponding locations or a strongly seasonal climate and vegetation in which the gymnosperms form the upper level of the vegetation, while the ferns concentrate on the more humid lower levels close to the floor. A decision on which of the two interpretations is more realistic can be made via a horizon-related evaluation considering the sedimentary facies. A statistical evaluation using correlation analyses of the frequency distributions of the various palynomorphs can also provide conclusive information here.

The palynoflora of the Huangshanjie Formation shows the highest diversity (58 species of 41 genera), both among the spore and among the seed plants. Interestingly, the gymnosperms (31 species of 20 genera) is dominant, with around 53% of the taxa, which may indicate a comparatively low humidity. Within the gymnosperms, both the conifers and the non-conifers play an important role. The main taxa are *Densoisporites velatus*, *Auritulinasporites intrastriatus*, *Annulispora folliculusa*, *Carnisporites granulatus*, *Duplexisporites problematicus*, *Kraeuselisporites reissingeri*, *Alisporites australis*, *Cycadopites reticulatus*, *Ovalipollis*, *Protopinus oberteticuspora*, etc. (Ashraf et al., 1998, 2001).

The palynoflora of the Haojiagou Formation still takes a diverse pattern in composition (52 species of 37 genera), among which the taxa *Lunatisporites rhaeticus*, *Calamospora rhaeticus*, *Chasmatosporites apertus*, *Duplexisporites problematicus*, and *Concavisporites* suggest the Late Triassic age for the Haojiagou Formation. A relative increase in spore taxa compared to the pollen taxa is observed in the transition from the Huangshanjie Formation to the Haojiagou Formation. Ecologically, the greater frequency of coaly layers in the Haojiagou Formation may indicate that the increased stress situation here was caused by water logging and also possibly nutrient starvation. Compared with the Huangshanjie Formation, the climate seems to have become more temperate and moist. According to the palynological findings (but also with the mega-plant fossil evidence), the Triassic-Jurassic (T/J) boundary is roughly to be placed at the boundary between the Haojiagou and Badaowan Formations. The palynoflora of the Lower Jurassic Badaowan Formation already looks significantly different from the Upper Haojiagou Formation. In the Badaowan Formation, the spores *Aratrisporites* and *Limatulasporites* have become very rare or are missing entirely; the same applies to the striate saccate gymnosperm pollen. An assignment of the Lower Jurassic Badaowan Formation is indicated by the taxa *Stereisporites perforatus*, *Duplexisporites problematicus*, *Concavisporites*, *Cyathidites*, and *Cerebropollenites*. According to this stratigraphic classification, the mass extinction or the decisive change in shape in the mainland vegetation would have taken place around the T/J boundary (Ashraf et al., 1998, 2001, 2010).

# 3.3 Three new species of Jurassic turtle *Xinjiangchelys*

Turtles are widely distributed in the Jurassic strata in Xinjiang, China, represented by the well-known genus *Xinjiangchelys* Yeh, 1986, belonging to the family Xinjiangchelidae (1990). Between 2001 and 2012, the authors' research team discovered three new species of turtles from the Jurassic strata in Xinjiang, including *Xinjiangchelys chowi* Matzke et al., 2005 from the Middle Jurassic Toutunhe Formation in Liuhuanggou near Urumqi ( Matzke et al., 2005), and *X. qiguensis* Matzke et al., 2004 from the Upper Jurassic Qigu Formation in Liuhuanggou (Matzke et al., 2004) (Fig. 73) which were both from the Junggar Basin; and *Xinjiangchelys wusu* Rabi et al., 2013 from the Qigu Formation in Shanshan of the Tu-Ha Basin (Rabi et al., 2013; Wings et al., 2012) (Figs. 28; 73).

## 3.3.1 *Xinjiangchelys chowi*

This taxon was found by Matzke et al. from the upper part of the Middle Jurassic (Callovian) Toutunhe Formation of the Liuhuanggou, about 40 km west of Urumqi in the Junggar Basin between 2001 and 2004, and the result was formally published in 2005 ( Fig. 73, B-D). The main diagnosis of *Xinjiangchelys chowi* is as follows:

Medium-sized eucryptodiran turtle with low, rounded shell; carapace at least 34.5 cm long; first neural largest and laterally expanded; dorsal articulation between seventh and eighth vertebra absent; first costal with free lateral rib end; lateral margin of second costal triangular; small peripheral fontanelles at least between first/second and second/third costals; two suprapygals, first clearly larger than second; short but markedly broadened pygal, with a medially thickened ridge on ventral side, extending onto second suprapygal; fifth vertebra significantly covering peripherals and pygal; anterior peripherals, including anteriormost part of sixth, guttered; twelfth marginals on pygal rather small; at least two sacral vertebrae, with only one sacral rib; plastron

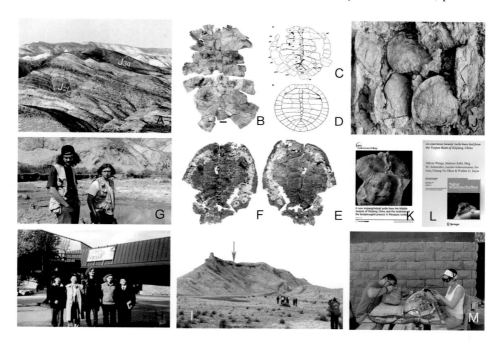

Fig. 73  Three new species of *Xinjiangchelys wusu* from Jurassic of Xinjiang

A. The turtle-bearing Middle Jurassic Toutunhe Formation and Upper Qigu Formation in Luhuanggou of Junggar Basin; B-D. *Xinjiangchelys chowi* from Toutunhe Formation, B. carapace, C. arrangement of elements of the carapace; D. reconstruction of the carapace (after Matzke et al., 2005); E, F. *X. qiguensis*, E. carapace in dorsal view, F. carapace in ventral view (after Maisch et al., 2004); G. The main discoverers of the above two new turtle species: Matzke A. (right), and Maisch M. (2004); H. Some members of the Sino-German research team in Stuttgart Museum (NH) in 2000, including Prof. Ye J. (IVPP, right 1), the nominator of the genus *Xinjiangchelys* ; I. The turtle site of *X. wusu* from Qiketai of Shanshan in Tu-Ha Basin; J-L. Skeletons of *X. wusu* and its publications (2013, 2015); M. Graduate Verena (right) preparing fossils in Shanshan of Xinjiang (2011)

loosely connected to carapace; large lateral fontanelles between carapace, hyo- and hypoplastron; posterior inguinal process of hypoplastron short, with long prominent lateral pegs, medial part of hypoplastron decreased in thickness, with pegs along median suture ( Fig. 73, C ) ( Matzke et al., 2005).

The holotype of *Xinjiangchelys chowi* consists of a nearly complete carapace and the right hypoplastron, which shows important autapomorphies of the family Xinjiangchelyidae and the genus *Xinjiangchelys*: first vertebra larger than second, anterior peripherals guttered, posterior peripherals expanded, a reduced first thoracic rib, and lateral pegs of the bridge of the plastron. It is distinguished from all other species of *Xinjiangchelys* by the following autapomorphies: a first costal rib with free rib end, at least two anterolateral peripheral fontanelles of the carapace, a thin plastron with large lateral fontanelles, and a median hypoplastral suture with strong pegs. These features indicate that *X. chowi* is the most derived known xinjiangchelyid turtle. Together with *X. tianshanensis*, this is the oldest xinjiangchelyid turtle known so far. Therefore, a long ghost lineage must be claimed for the Xinjiangchelyidae (Matzke et al., 2005).

### 3.3.2 *Xinjiangchelys qiguensis*

*Xinjiangchelys qiguensis* Matzke et al., 2004 was first found in the Upper Jurassic Qigu Formation in the Liuhuanggou section near Urumqi in the Junggar Basin by Matzke A. and his research group of the authors' research team (Matzke et al., 2004). This taxon is the most plesiomorphic species of *Xinjiangchelys*, and has nearly the same carapace length as *Xinjiangchelys chowi*. The holotype (SGP 2000/5) of *Xinjiangchelys qiguensis* consists of a partial skeleton with carapace, plastron, both scapulae, pelvis, ulna, and almost all cervical vertebrae (Matzke et al., 2004).

*Xinjiangchelys qiguensis* has been diagnosed as follows (Matzke et al., 2004): Peripherals 1-6 guttered (potential autapomorphy); increased width of the first vertebral scute compared with the width of the nuchal (potential autapomorphy), pygal large with emarginated anterior edge; the first and fifth vertebrals extend onto the peripherals (potential autapomorphy); three pairs of gulars (potential autapomorphy); intergular without contact to humeral (potential autapomorphy); axillary scute enlarged and first inframarginal reduced in size (potential autapomorphy); neck longer than half of the carapace length, with elongated amphicoelous cervical vertebrae; all neural spines low, ventral keel increasing from the anterior to the posterior vertebrae; centrum of first thoracic vertebra anteriorly oriented; two sacral vertebrae present, only the first bears a sacral rib; scapula with long acromial process and shorter scapular process, anteromedial tip of acromial process curves ventrally with small ventral projection, angle between both processes about 100 degrees; broad pelvis with short ilium, posterior iliac process longer than iliac shaft (potential autapomorphy), enclosing an angle of 90 degrees with iliac shaft, ilium with long anteromedial projection, and two separate thyroid fenestrae (potential autapomorphy).

The neurals of *Xinjiangchelys qiguensis* are all somewhat asymmetric and no gap between the 7$^{th}$ and 8$^{th}$ neurals. In anterolateral part of the carapace, no free rib ends and no fontanelles are found. The posteriormost margin of the carapace is anteriorly curved and not posteriorly pointed as in *Xinjiangchelys chowi*. The articular facet for the pelvis, exposed on the ventromedial part of the 8$^{th}$ costals, is markedly smaller in *X. chowi* than *X. qiguensis*. In contrast to that *X. chowi*, the plastron of *X. qiguensis* is strongly sutured to the carapace and no lateral fontanelles or medial pegs between the hypoplastra are observable. The thickness of the medial parts of the hypoplastra of *X. qiguensis* does not decrease, as the case of *X. chowi* (Maich et al., 2004)(Figs. 73).

### 3.3.3 *Xinjiangchelys wusu*

*Xinjiangchelys wusu* Rabi et al., 2013 was found by the authors' research team at the top of a hill in the Gobi of Qiketai, eastern Shanshan (42°57'31.5"N, 90°33'21.0"E), about 7 km northeast of the dinosaur track fossil site and 5 km northwest of the dinosaur *Xinjiangotitan* site (Fig. 28, B). The turtle-bearing beds are composed of grayish green calcareous sandstones and siltstones sandwiched in the brick red and variegated sandstones of the Upper Jurassic Qigu Formation (BGMRX, 1993). A mass accumulation of turtle carapaces was discovered, with about 36 turtles found per m$^2$, which comprised also three nearly complete skeletons (Figs. 28, C, D; 29, A-E; 73, J). The complete turtle shells are about 20 cm long. It had been estimated that about 720 turtle specimens are preserved in an area of 20 m$^2$ (Rabi et al., 2013).

The main characteristics of *X. wusu* are as follows: the skull wide; palatine and frontal more widely involved in orbital forming; prefrontal completely separated by the frontal; pterygoid space developing; and the vertebral shield narrow and long (Figs. 28; 29). More importantly, this species develops some primitive skull base features, such as pterygoid space and basal sphenoid process, which are different from the highly ossified skull base of modern turtles, indicating the transformation of ecological adaptation in the evolution of turtles. At the same time, these characteristics also provide a base for exploring the evolution of the carotid system of turtles. The discovery of complete skeletons of *X. wusu* is most important for the understanding of skeletal evolution in Mesozoic turtles (Rabi et al., 2013; Wings et al., 2012).

The mass accumulation of *Xinjiangchelys wusu* carapaces in the Shanshan area is likely due to seasonal droughts in the early Late Jurassic stage, which caused a gathering of turtles in shrinking ponds. Their paleo-environment in the lower Upper Jurassic of the Shanshan area in the Tu-Ha Basin appears to be similar to that of certain extant ecosystems, e.g. of Venezuela in South America (Figs. 29, G) or of the Murray River in Australia.

# 3.4  Middle Jurassic flora from Baiyanghe

## 3.4.1  Baiyanghe flora

The Baiyanghe flora has been found in the coal-bearing strata of the Middle Jurassic Xishanyao Formation in the Baiyang River valley (84°56′E, 46°22′N in central coordinate) of northeastern Tuoli County, northwestern Junggar Basin. Influenced by the field work conditions at that time, the plant fossils collected by Obruchev were limited in number, which let Seward (1911) report only 21 species belonging to 16 genera, including: *Equisetites ferganensis* Sternberg, *Coniopteris hymenophylloides* Brongniart, *C. quinqueloba* (Phillips) Seward, *Eboracia lobifolia* (Phill.) Thomas, *Sphenopteris modesta* Leckenby, *Raphaelia diamensis* Seward, *Cladophlebis* sp., *Ginkgo obrutschewi* Seward, *Baiera lindleyana* Schimper, *Phoenicopsis angustifolia* f. *media* Krasser, *Podozamites lanceolatus* (L. et H.) Braun, *Podozamites* cf. *pulchellus* Heer, *Pityophyllum* cf. *staratschini* Heer, and *?Sphenolepidium* sp., etc. To further study the Baiyanghe flora and to check on the fossil site, Sun G. visited the Russian Geological Museum (VSEGEI) in St. Petersburg, several times, both to investigate Obruchev's collection housed in the museum, and check the original Geological Map of Junggar made by Obruchev in 1914, which documents the main fossil plant site in the Ak-djar valley where he worked, and which is now named the Baiyang River valley (Fig. 30).

So far, more than 46 species of 28 genera of plant fossils have been found in this site, and have been studied in an overall manner by the authors, with the main achievements including (1) newly expanding the cognition of the composition of the flora, such as confirming the existence of *Thallites*, *Neocalamites*, *Ferganella*, *Ixostrobus*, *Czekanowskia* (*Vachramee*) *baikalica*, and *Leptostrobus laxifolia* for the first time; (2) finding of the new taxa *Sphenobaiera jungarensis* Sun et Miao and *Scarburgia baiyangheensis* Yang et al., etc. (Sun et al., 2006; Yang et al., 2017); (3) revising or combining several important taxa reported by Seward (1911), such as *Equisetites ferganensis*, *Coniopteris hymenophylloides*, *C. quinqueloba*, *Raphaelia diamensis*, *Eboracia lobifolia*, and *Sphenopteris modesta*, etc. (Sun et al., 2001b, 2010, 2021; Miao, 2005, 2006, 2017; Sun et al., 2006; Yang et al., 2017, 2018); and (4) a detailed cuticular study of *Ginkgo obrutschewi* and a detailed study of the reproductive organs *in situ* of *Coniopteris hymenophylloides* (Sun et al., 2010, 2021).

At present, the following plant fossils have been identified in the Baiyanghe flora (Figs. 31-33):

**Bryophytes**

1. *Thallites* sp.

**Equisetales**

2. *Equisetites ferganensis* Seward

3. *Equisetites* sp.

4. *Neocalamites hoerensis* (Schimp.) Halle

**Filicinae**

5. *Coniopteris burejensis* (Zal.) Seward

6. *C. hymenophylloides* (Brongn.) Harris

7. *Eboracia lobifolia* (Phill.) Thomas

8. *Cladophlebis shansiensis* Sze

9. *C. tsaidamensis* Sze

10. *Cladophlebis* sp.

11. *Raphaelia diamensis* Seward

12. *Raphaelia* sp.

**Bennettitales**

13. *Nilssoniopteris* sp.

**Czekanowsiales**

14. *Phoenicopsis angustifolia* Heer

15. *P. speciosa* Heer

16. *Czekanowskia rigida* Heer

17. *C. (Vachrameevia) baikalica* Kirichkova et Samylina

18. *Leptostrobus laxifolia* Heer

19. *Leptostrobus* sp.

20. *Ixostrobus* sp.

## Ginkgoales

21. *Baiera lindleyana* Schimper

22. *Ginkgo obrutschewi* Seward

23. *G. huttonii* Harris?

24. *Ginkgo* sp.

25. *Sphenobaiera jungarensis* Sun et Miao

26. *Stenorachis lepida* (Heer) Seward

## Coniferales

27. *Pityophyllum* sp.

28. *Schizolepis* sp.

29. *Cupressinocladus?* sp.

30. *Taxodium?* sp.

31. *Lindlecladus laceolatus* (L. et H.) Harris

32. *Ferganella* sp.

33. *Scarburgia baiyangheensis* Yang et al.

34. *Elatides ovalis* Heer?

35. *Elatocladus minutus* Doludenko

36. *Elatocladus pinnata* Sun et Zheng

37-38. *Elatocladus* spp. 1, 2

39-40. *Strobilites* spp. 1, 2

## Unclassified taxa and fossil woods

41-45. *Carpolithus* spp. 1-5

46. Fossil wood (in study)

The Baiyanghe flora is characterized by the *Coniopteris-Phoenicopsis-Ginkgo* assemblage, and generally belongs to the late assemblage of the *Coniopteris-Phoenicopsis* flora in Northern China, showing an early Middle Jurassic age (Zhou, 1995; Sun et al., 2021). The Baiyanghe flora mainly reflects a warm humid climate of the warm temperate to the temperate zone with seasonal changes during the Middle Jurassic in Xinjiang.

## 3.4.2 On *Ginkgo obrutschewi*

*Ginkgo obrutschewi* was found in the Baiyang River valley and named by Seward (1911). Because of its wide distribution in the Middle Jurassic, the paleobotanists have paid much attention to its research. However, the original material of *G. obrutschewi* consisted only of three incompletely preserved specimens and the original description was based only on their external features and limited cuticular characteristics by a light microscope (LM) (Seward, 1911).

In the past ten years, the authors have collected a number of new fossils of *G. obrutschewi* in the topotype site, the Baiyang River valley section, and studied in detail its epidermal structures by SEM, which revealed the following main characteristics of its cuticles: (1) basically hypostomatic in type; (2) the ordinary epidermal cells are papillate or domed cutinizing on periclinal walls; (3) the anticlinal walls straight or slightly curved; and (4) stomata haplocheilic, with guard cells strongly lipped cutinized thickening in inner edges, and the subsidiary cells usually strongly cutinized.

In the upper cuticles of the leaves, the ordinary epidermal cells are usually irregularly polygonal in shape, with the anticlinal walls cuticularly thickening and slightly curved. In the lower cuticles, the non-stomatal zone is composed of 9-14 rows of long rectangular ordinary epidermal cells, and the stomatal zone is usually composed of nearly equiaxed polygonal ordinary epidermal cells. The stomata are usually not in orientation, their guard cells cuticularly thickening with radially cuticular striations in inner edges. The subsidiary cells are 4-5 in number, usually strongly papillate. The stomatal density (SD) is 31-36 /mm$^2$. (Miao, 2006, 2017; Sun et al., 2004, 2010a, 2021 ) (Fig. 34, Q–V).

In 2011, a significant research work related to the study of *G. obrutschewi* was reported by Nosova (BI RAS) et al. They studied the cuticles of the type specimen of *G. obrutschewi* (No. 29 of Seward, 1911) and of a topotype specimen (No. 31 of Seward, 1911), housed in the Russian Geological Museum (VSEGEI) , and reported the cuticles of *G. obrutschewi* being of amphistomatic type (Nosova et al., 2011). Although there are some different opinions on this report (Miao, 2017), this new finding of Nosova et al. (2011) is very significant for the further study of *G. obrutschewi*. Now, Sun G. and Dr. Nosova N. have conducted some new cooperative work on the Late Cretaceous flora from NE China (Fig. 34, X).

## 3.5  Jurassic mammal assemblage from Liuhuanggou

Jurassic mammals have been discovered in several localities in Xinjiang. Liuhuanggou, about 50 km west of Urumqi, is the main site in the Junggar Basin, yielding abundant mammals from the Upper Jurassic Qigu Formation. The Liuhuanggou mammal assemblage was first discovered by the research group headed by Pfretzschner H.-U. and Martin T. of the authors' research team and exploited from 2001 to 2010. The mammals are represented by *Dsungarodon-Tegotherium-Sineleutherus-Nanolestes* assemblage, which is dominated by docodontans, representing the highest diversity in the Late Jurassic of Asia (Pfretzschner et al., 2005; Martin et al., 2010b) (Figs. 35; 74). The mammal fossils are mainly preserved as jaw fragments and isolated teeth and comprise the following five taxa:

### Docodonta

1. *Dsungarodon zuoi* Pfretzschner et Martin, 2005

2. *Tegotherium* sp.

### Haramiyida

3. *Sineleutherus uyguricus* Martin et al., 2010

### Stem-Zatheria

4. *Nanolestes mackennai* Martin et al., 2010

### Eutriconodonta

5. Eutriconodonta indet.

## (1) *Dsungarodon zuoi* Pfretzschner et Martin, 2005 ( Fig. 74, E )

*Dsungarodon* and its type-species *D. zuoi* were first found in the Qigu Formation in Liuhuanggou by the authors' research team and named by Pfretzschner and Martin (Pfretzschner et al., 2005). The taxon belongs to the Docodontidae and the generic diagnosis was revised by Martin et al. (2010) as follows: "Differs from all other docodontans by a large pseudotalonid basin bordered by crests a-b, a-g, and b-g, crenulations on the distal side of the lower molar crowns, and an additional groove above the Meckelian groove separated from the more posterior trough for the postdentary bones." Compared with *Simpsonodon* in the structure of the mesial cingulid on the lower molars, *Dsungarodon* has two parts, a robust labial rim pointing mesiolingually and a weaker lingual rim extending to the base of cusp g; the two rims meet mesially at an acute angle, while *Simpsonodon* has a continuous mesial cingulid wrapping around the mesial end of the crown between cusps b and g, with a rounded mesial end. In addition, *Dsungarodon* possesses a weakly developed cusp c on the ultimate lower molar, while it is completely absent in *Simpsonodon*. The genus *Acuodulodon* Hu et al. 2007 from the Upper Jurassic Shishugou Formation in Wucaiwan in the central Junggar Basin (Hu et al., 2007) is extremely similar to

*Dsungarodon* and the authors have considered it synonymous with *Dsungarodon*. The specific name of the type-species, *-zuoi*, has been awarded to Prof. Zuo X. Y., the former Director of the Regional Geological Survey of Xinjiang No. 1, to thank him for his support of the SGWSGX (Martin et al., 2010b). Docodonta were first found in the Jurassic of the United States and Europe, and later in Central Asia (Kyrgyzstan and Russia), mostly represented by jaws and teeth. However, in 2019, one of the authors, Prof. Martin co-authored the description and analysis of a complete skeleton of the small docodontan (ca. 15 cm long) *Microdocodon* from the Upper Jurassic of eastern Inner Mongolia, China (Zhou et al., 2019).

## (2) *Tegotherium* sp. ( Fig. 74, G )

This genus belongs to the docodontan family Tegotheriidae, and was established by Tatarinov (1994) based on a molar from the Upper Jurassic of Shar Teeg in Mongolia. The main difference between this genus and the Docodontidae is a different arrangement of the cusps around the pseudotalonid basin which is bordered by crests a-g, a-b, b-e, and g-e in Tegotheriidae (Fig. 74, G). Among tegotheriids, *Tegotherium* differs from *Tashkumyrodon* Martin and Averianov, 2004 by a strong a-d crest and lack of a c-d crest (in *Tashkumyrodon*, the c-d crest is strong, and crest a-d is absent); it differs from *Krusatodon* Sigogneau-Russell, 2003 by the lack of additional crests in the distal portion of the crown of the lower molars; it differs from both *Tashkumyrodon* and *Krusatodon* by a complete lingual cingulid; it differs from *Sibirotherium* Maschenko et al., 2003 by a stronger crest e-g and much weaker distal cingulid (crest d-dd). Further, it differs from *Krusatodon* and *Sibirotherium* by the presence of two (X and Y) rather than three (X, Y, and Z) lingual cusps on the upper molars (Martin et al., 2010b).

## (3) *Sineleutherus uyguricus* Martin, Averianov et Pfretzschner, 2010 ( Fig. 74, H)

This new taxon was found and established by one of the authors and his group in 2010

Fig. 74  Late Jurassic mammal fauna from Liuhuanggou of Junggar, Xinjiang

A.  Liuhuanggou; B. Fossil site of Qigu Formation ($J_3$) in the Liuhuanggou; C. Prof. Pfretzschner H.-U.; D. Prof. Martin T.; E. *Dsungarodon zuoi*; F. Prof. Zuo X. Y.; G. *Tegotherium* sp.; H. *Sineleutherus uyguricus*; I, *Nanolestes mackennai*; J. Eutriconodonta indet.; K-P. Field work at Liuhuanggou, including the fossil collections, transportation, washing and selections; K showing Prof. Pfretzschner H.-U. and O showing Prof. Martin T. (2005) (E-J after Martin et al., 2010b)

(Martin et al., 2010b). This genus belongs to the allotherian Eleutherodontidae, and differs from other allotherians in exhibiting ovoid molariform teeth, with two rows of cusps that are continuously around the distal end and placement of the largest cusp on the lower molariforms at the mesial end of the labial row. Compared with Multituberculata, the present taxon retains an extensive orthal component of occlusion. *Sineleutherus* differs from *Eleutherodon* Kermack et al., 1998, by larger and fewer marginal cusps on the lower molariform teeth and the absence of numerous transverse ridges (crenulations, or "fluting") in the central basin (Martin et al., 2010b).

## (4) *Nanolestes mackennai* Martin, Averianov et Pfretzschner, 2010 ( Fig. 74, I)

This new species was found and named by one of the authors (Martin T.) and his group from the Upper Jurassic Qigu Formation in Liuhuanggou in 2010 ( Martin et al., 2010b). The genus *Nanolestes* was established by Martin (2002) and belongs to stem-Zatheria. The new specific name was accredited to Dr. M. C. McKenna for his contributions to the study of Mesozoic mammals. The original generic diagnosis of *Nanolestes* was made by Martin in 2002, and amended in 2010 after the discovery of the new species *N. mackennai* (Martin et al., 2010b). This species is characterized by a comparatively short trigonid (angle 50° ) and a metaconid which is in line with the protoconid, not shifted posteriorly in molars distal to m1. (after reinterpretation of Gui Mam 1005 as m1, and not as a deciduous premolar as interpreted by Lopatin and Averianov, 2006).

## (5) Eutriconodonta indet. ( Fig. 74, J)

The tooth is assigned to the Eutriconodonta, but cannot be assigned at the specific level due to its fragmentary nature. The main characteristics of the tooth are a large main cusp a, somewhat curved distally and missing the apex; a comparatively large distal cusp c of about half the height, sharp, with dorsally directed apex; and a small but distinct more distal cusp d, not cingular and well separated from the cingulid. The anterior side of the crown is missing. The distal end of the crown is sharply pointed, and there is no distinct distal cingulid, while the lingual and labial cingulids extend to the pointed distal end of the tooth. On both sides, the cingulids are confined to the posterior part of the crown and do not extend mesially beyond cusp c. The lingual cingulid is about two fold longer than the labial cingulid. The crown is slightly asymmetrical on the posterior side in occlusal view, with the lingual side more expanded, and more symmetrical on the anterior side. The enamel surface is sculptured by very fine striations. The tooth was double-rooted, with a large space between the roots; however, the roots are not preserved (Martin et al., 2010b) .

The Liuhuanggou Jurassic mammal assemblage is the most diverse of the Late Jurassic in Asia and has something in common with the Shar Teeg site in Mongolia the docodontan genus *Tegotherium* (*T. gubini* Tatarinov, 1994). The Upper Shaximiao Formation of Shilongzhai, Sichuan of China that yielded *Shuotherium* is most likely Middle Jurassic in age, because *Shuotherium* has been recorded in the well-dated Bathonian of Kirtlington in England. However, the Liuhuanggou mammal assemblage also shares some similarities with the well-known Bathonian Forest Marble mammal assemblage of England, such as the presence of eleutherodontid allotherians, simpsonodontid and tegotheriid docodontans, as well as stem-lineage zatherians. The Liuhuanggou mammal assemblage appears essentially Middle Jurassic in composition, and differs clearly from the Early Cretaceous mammal assemblages of Asia which are dominated by multituberculates, gobiconodontid eutriconodontans, "symmetrodontans" , and eutherians (Martin et al., 2010b).

## 3.6  Jurassic dinosaur tracks in Shanshan

The Shanshan dinosaur tracks were found in the Middle Jurassic Sanjianfang Formation in the Gobi, south of Qiketai, eastern Shanshan County. The formation consists of grayish green and yellowish green sandstone and mudstone intercalated with purple sandstone and siltstone. The formation shows fluvial-lacustrine facies, and measures about 400 m in total thickness. About 155 dinosaur tracks in total are preserved as natural casts on the underside of an overturned bank of sandstone extending more than 100 m in NE-SW direction (Figs. 36; 37; 75). The tracks are densely crowded and irregularly distributed. Due to rapid erosion of the soft greenish to purplish mudstone, the original track imprints are not preserved.

The Shanshan dinosaur tracks are all tridactyl mesaxonic in type, with a pointed claw in front of each toe. Some tracks show well-defined phalangeal pads, heels, and, in some cases, an indistinct impression of the distal part of the metatarsus. One footprint shows prominent retro-scratches. They all belong to carnivorous theropod dinosaurs. Based mainly on their general features, the dinosaur tracks can be divided into two distinct morphotypes. One type (type A) is larger, up to 33 cm long, and can be assigned to a large carnivorous dinosaur, while the other type (type B) is small and slender, possibly belonging to a skinny coelurosaurid (Wings et al., 2007).

The dinosaur tracks of Shanshan are described as follows.

**Morphotype A:** These tracks are longer than wide and generally of deltoid shape. The total width ranges from 17.5 cm to 38.2 cm. A heel is more or less well defined. Digit III is the longest, with a length of 18.3-33.6 cm. Digits II and IV are about 25% shorter than digit III, with digit II tending to be slightly longer than digit IV. Phalangeal pads are moderately well defined; four pads can be found on digit III, with the proximal-most being confluent with the heel; two pads can be distinguished on digit II and three pads on digit IV. The pads tend to be as long as broad or slightly longer than broad, separated by relatively shallow constrictions, which gives this footprint morphotype a massive appearance. Digit III is the broadest, attaining its maximum width, ranging from 3.9 cm to 12.1 cm in its distal half. The average angle between digits II and III is 37° and between digits III and IV, 40°, resulting in a total divergence of approximately 77°. Digits II and III often have their tips deflected medially (away from digit Ⅳ). The interdigital area is more deeply indented between Ⅱ and Ⅲ than between Ⅲ and Ⅳ. Distinctive V-shaped claw impressions mark the distal tip, especially of digits Ⅱ and Ⅲ ( Fig. 37, A).

**Morphotype B:** The tracks are elongate and have a slender and gracile appearance. The total width of the footprints is 12.2-33.3 cm. A heel is present but weakly defined. Digit Ⅲ is the longest, attaining a length of 11.4-29.2 cm. Digits Ⅱ and Ⅳ are subequal in length and approximately 30% shorter than digit Ⅲ. Three well-defined phalangeal pads can be found on digit Ⅲ, with a possible fourth pad being part of the heel, two pads on digit Ⅱ and three on digit Ⅳ. The phalangeal pads are elongate. Especially on digit Ⅲ, their width decreases toward the distal tip of the digit, the broadest part of the digit (2.8-7.2 cm) being in its proximal half. The average divergence between digits Ⅱ and Ⅳ is 73°, with subequal angles between digits Ⅱ and Ⅲ and digits Ⅲ and Ⅳ. No clear difference between both interdigital areas is observed. Well-defined V-shaped and pointed claw impressions can be found at the tip of the digits (Fig. 37, B). The claw impressions tend to be smaller than those of morphotype A.

Xing et al. (2014) re-studied the dinosaur tracks from Shanshan, and suggested that the Shanshan dinosaur tracks belong to a single ichnotaxon, *Changpeipus carbonicusand* and that the differences in size and shape observed by Wings et al. (2007) are due to differing individual age and substrate conditions (Xing et al., 2014). According to Xing et al. (2014), the presence of two different biological species of track-making dinosaurs cannot be completely ruled out, but appears unlikely.

In addition, in the upper horizons of the Sanjianfang Formation, the authors' team also found invertebrate trace fossils, sauropod dinosaur bones, and fossil wood (Fig. 75, H and I). The tracks and other traces are preserved in sediments of fluvial-lacustrine facies. The paleoecological analysis of these traces suggests that the dinosaurs lived in a lakeside swamp environment not far from forested areas in the

Tu-Ha Basin. The climate during the Middle Jurassic was humid and warm. Xing et al. (2014) suggested that the abundant invertebrate trace fossils indicate a gradually expanding and deepening lake, and that the invertebrate trace fossil *Lockeia* found in this site should be classified as *Fuersichnus* .

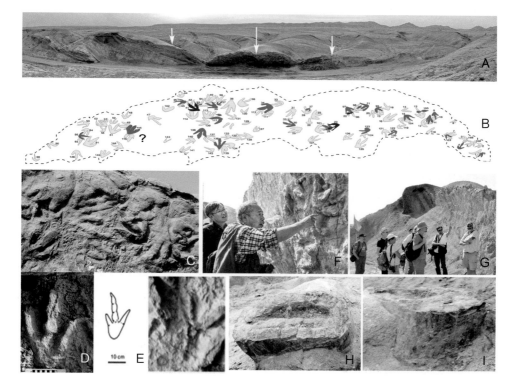

Fig. 75  Discovery of Jurassic dinosaur tracks in Shanshan

A, B. Outcrop of dinosaur tracks and their illustration; C-E. Dinosaur tracks: D showing the morphotype A, E showing the morphotype B and its reconstruction; F, G. Prof. Dong (F) and Dr. Wings (G) explaining the dinosaur tracks in Shanshan (2013); H, I. Associated fossils: H showing a dinosaur bone, I showing a fossil wood

# 3.7 *Xinjiangotitan*, the largest known dinosaur in Xinjiang

## 3.7.1 *Xinjiangotitan shanshanensis*

*Xinjiangotitan* is the best preserved and largest known Jurassic sauropod dinosaur from Xinjiang, and was found in the Upper Jurassic Qigu Formation south of Qiketai, eastern Shanshan. The formation is composed of variegated sandstone, siltstone, and fine conglomerate with mudstone. The articulated dinosaur skeleton was discovered and collected by the authors' research team in yellowish purple sandstone between 2011 and 2012, and formally described in the journal *Global Geology* in 2013, with the type-species *X. shanshanensis*. The paper on the discovery of *Xinjiangotitan* was first published by the members of the authors' research team, including Wu, Wings, Zhou, Sekiya, and Dong (Wu et al., 2013) (Fig. 76).

Until June 2013, the authors' research team had found the following articulated bones of *Xinjiangotitan shanshanensis*: three cervical vertebrae, 12 dorsal vertebrae, five sacral vertebrae, the pelvic girdle consisting of ischium, ilium, and pubis, the left femur (ca. 1.6 m long), tibia and fibula, etc. According to the size of these bones, the length of this dinosaur was estimated at 30 m, which is the largest known Jurassic dinosaur in China including Xinjiang. Besides, this dinosaur is one of the best preserved of China (Fig. 34). In their phylogenetic analysis, Wu et al. (2013) considered *Xinjiangotitan* as the sister taxon of *Mamenchisaurus*, and both taxa belong to the sauropod family Mamenchisauridae Young et Chao, 1972. (Fig. 39).

Between 2016 and 2018, Li D. Q., a Chinese expert on dinosaurs, was entrusted by the authors to conduct a follow-up excavation of *Xinjiangotitan*, and made important progress with his excavation group. With the new findings the partial skeleton now comprises parts of the skull, 18 cervical vertebrae, 12 almost complete dorsal vertebrae, five articulated sacral vertebrae, 39 caudal vertebrae,

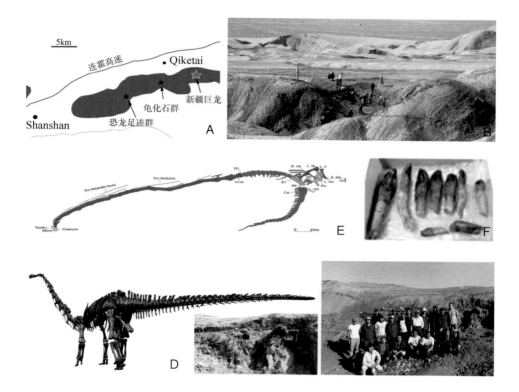

Fig. 76  *Xinjiangotitan shanshanensis* Wu et al., 2013

A. Sketch illustration on the geographic position of *Xinjiangotitan* (red star); B. Excavation view; C. Skeleton of *Xinjiangotitan in situ*; D. Reconstructive skeleton of *Xinjiangotitan* (after Bureau of Natural Resources, Shanshan, 2016); E. Illustration of *Xinjiangotitan* (E, after Zhang et al.,2018) ; F. The teeth of *Xiniangotitan*; G. Members of the Sino-German team working for the excavations in Shanshan (2012)

partial cervical, and dorsal ribs, 18 chevrons, the left ilium, both pubes, both ischia, the left femur, the left tibia, the left fibula, the left astragalus, and a partial left pes. Based on these new discoveries, Li and his research group were able to describe *X. shanshanensis* in more detail (Zhang et al., 2018). *Xinjiangotitan* is characterized, according to the revised diagnosis by Zhang et al., (2018) by the following combination of characteristics: Ventral keel extending almost along the middle-posterior portion of the ventral surface in cv 3-12 (confined to the anterior concavity in cv 13-17; longest cervical is cv 12; the ratio of the length of cv 3 ： cv 2 is 1.58; lateral longitudinal pleurocoel divided by an antero-dorsally oblique ridge into an anterior and a posterior part; horizontal ridge present dorsal to the lateral pneumatic fossa in cv 9-18; pneumatopores in spinodiapophyseal fossa in the middle-posterior cervical vertebrae; only the last cervical neural spine bifurcated with the median tubercle.

### 3.7.2 New discoveries of other related fossils

Besides, the authors' research team has found a large theropod tooth in the *Xinjiangotitan* locality (Fig. 38, E). In addition, near the fossil site (East Point No. 1, 42 °57'47.6"N, 90 °34'32.5"E), the authors discovered eight associated cervical vertebrae, about 16 m long in total (Fig. 41, H).

It is worth mentioning that in the Qiketai Formation, which underlies the Qigu Formation yielding *Xinjiangotitan*, the authors have also found some fossils such as fish scales, gastropods, and bivalves including *Kajia ovalis* (Martinson) and *Kijia kweizhouensis* (Grabau) identified by Sha J. G. which had been recorded earlier from the Middle Jurassic Chakmak and Kizilnur formations in Kuqa area, Tarim Basin (Fig. 41, A-G). These bivalve fossils appear to provide indirect evidence for an early Late Jurassic age of the Qigu Formation.

# 3.8 Middle Jurassic forest of Shaerhu

The Shaherhu coalfield is located in the middle and southern Tu-Ha Basin (central coordinates 42°35′N, 91°20′E ), about 130 km east of Shanshan County, and about 160 km west of Hami City.

The coal-bearing Middle Jurassic Xishanyao Formation of Shaerhu is mainly composed of sandstone, siltstone, and carbonaceous mudstone, rich in thick coal seams and yielding abundant plant fossils. The RGSX No.1 mapped this area in 1989, and measured the geological section of the formation (BGMRX, 1993), and the authors have undertaken comprehensive geological and paleontological studies of this formation since 2008 (Dong et al., 2011) (Fig. 42, B-G).

The stratigraphic sequence of the Xishanyao Formation in the Shaerhu coalfield is as follows:

**Overlying:** Toutunhe Formation ($J_{2t}$): brick red fine sandstones with mudstones and siltstones

---------------------conformably---------------------

**Xishanyao Formation ($J_{2x}$):**                                                                   thickness (unit: m)

4. The upper part consists of gray-green mudstone, siltstone intercalated with carbonaceous mudstone and coal seams; the lower part consists of grayish green siltstone and medium fine grained sandstone--------------- 143.59

3. The upper part consists of dark gray mudstone, sandstone, carbonaceous mudstone and coal seams; the lower part consists of gray argillaceous siltstone intercalated with carbonaceous mudstone and coal seams------------ 93.92

2. Dark gray, light gray argillaceous siltstone, carbonaceous mudstone, and coal seams with silty mudstone ------------------------------------------ 85.36

1. Gray siltstones, argillaceous siltstone, sandstone, carbonaceous sandstone and coal seams; producing a lot of plant fossils --------------------------- 185.59

--------------------unconformably--------------------

**Underlying:** Aqikebulake Formation ($P_{2a}$):

Gray-green basalts, andesites, tuffaceous sandstone, and breccias

## 3.8.1 Mega-plant fossils

Dong et al. of the authors have collected a lot of fossils including 227 megaplant remains and 35 samples of sporopollen, which mainly derive from the siltstone and coal seams in the middle and lower members of the Xishanyao Formation in the Shaerhu coalfield. The megaflora is composed of over 20 species of 18 genera and is characterized by cycads and ferns, while the ginkgos, czekanowskialeans, and conifers also cover a certain portion. Among them, the Equisetales are represented by two species of two genera (10% of total species) with a new species (*Equisetites shaerhuensis* Dong); the Filicinae by six species of four genera (30%); the Cycadophytes (Bennettitales) by six species of four genera (30%); the Czeknowskiales and Ginkgoales by five species of four genera (20%), and the Coniferales by two species of two genera (10%) (Dong et al., 2011).

The major taxa of the megaflora are the following (Figs. 43-45):

### Equisetales

1. *Equisetites shaerhuensis* Dong

2. *Annulariopsis simpsoni* (Phill.) Harris

### Filicinae

3. *Coniopteris hymenophylloides* Brongniart

4. *Osmundopsis sturi* ( Raciborski ) Harris

5. *Cladophlebis whitbiensis* Harris

6. *Cladophlebis* cf. *delicatula* Yabe et Oishi

7. *Raphaelia diamensis* Seward

### Bennettitales

8. *Pterophyllum propinquum* Goeppert

9. *Nilssoniopteris* sp.

## Cycadales

10. *Nilssonia* cf. *acuminata* Presl

11. *Nilssonia* cf. *tenuinervis* Nathorst

12. *Ctenis* cf. *kaneharai* Yokoyama

## Ginkgoales

13. *Ginkgo digitata* (Brongn.) Heer

14. *Ginkgo obrutschewi* Seward

15. *Ginkgoites* cf. *lepidus* (Heer) Florin

## Czekanowskiales

16. *Phoenicopsis angustifolia* Heer

17. *Czekanowskia* sp.

## Coniferales

18. *Lindleycladus lanceolatus* (L. et H.) Harris

19. *Strobilites* sp.

## Unclassified taxa

20. *Carpolithus* sp.

Among the Equisetales, the new species *Equisetites shaerhuensis* Dong is characterized by its sheaths (Dong, 2012) (Fig. 43, A-E, I, and J). For a detailed understanding of this species, its features are further summarized as follows.

### *Equisetites shaerhuensis* Dong (Fig. 44)

(cf. Dong, 2010, p. 28, pl. 1 figs. 1-3, 6-8; pl. 2, figs. 1-2)

Diagnosis (in sum): Stem wide line-formed, noded, internode more than 1 cm long and 3.5-4 mm wide, with longitudinal ridges and groves alternating on the stem surface. Nodes with slender leaves, about 17 in number for each node. Leaves 1.5-2.1 cm long, 0.5 mm wide, lanceolate in form, with an obvious midvein. The basal lower

1/2-1/3 of leaves with fused sheath, 3-4 mm long and 0.6-0.7 mm wide, with very thin longitudinal striations on the surface of the base of the sheath. The epidermal cells of the sheath elongated and irregular quadrilateral in form, 33-166 μm × 20-50 μm in size. Diaphragms elliptic in form, with about 17 radiating lines, projecting from the center of diaphragma. This species is different from the known species of *Equisetites* in its unique leaf sheaths, and diaphragms (Fig. 44).

Among the Filicinae, *Coniopteris hymenophylloides* is dominant, with pinnules reduced as a thin stalk in shape. Its sporangia are rounded in form, about 227 μm × 172 μm in size, with the annulus about 33 μm wide, and terminating on the top of *Sphenopteris*-leaf typed veins (Fig. 43, O-R). Besides, the authors found *Osmundopsis sturi* which is suggested as the fertile pinnae of *Cladophlebis*. Associated with the two taxa mentioned above, the ferns include also *Cladophlebis whitbiensis*, *Cladophlebis* cf. *delicatula*, and *Raphaelia diamensis* (Fig. 43, T-W).

In the megaflora, cycadaleans and bennettitaleans are also abundant (Fig. 45, A-G). The Bennettitales include *Pterophyllum propinquum*, *Anomozamites* cf. *minor*, *Nilssoniopteris* sp., and *Taeniopteris* sp., while the Cycadales include *Nilssonia* cf. *acuminata*, *Nilssonia* cf. *tenuinervis* and *Ctenis* cf. *kaneharai*, etc. Ginkgos and czekanowskialeans are also abundant (Fig. 45, H-M). Ginkgoales mainly include *Ginkgo digitata*, *G. obrutschewi*, *Ginkgoites* cf. *lepidus*, etc. The Czekanowskiales mainly include *Phoenicopsis angustifolia* and *Czekanowskia* sp. Conifers are rare in the floral composition and are mainly represented by *Lindleycladus lanceolatus*, etc. (Fig. 45, N).

## 3.8.2 Sporopollen fossils

Besides the mega-plant fossils, the Shaerhu flora contains also abundant sporopollen fossils. The palynoflora is composed of over 49 species of 34 genera, in which the fern spores account for ca. 43%, and the gymnosperm pollen accounts for ca.

57%. The spore fossils are represented by the *Neoraistrickia-Lycopodiumsporites-Alsophilidites-Deltoidospora* assemblage (Fig. 46), while the pollen is represented by the *Cycadopites-Abietineaepollenites-Podocarpidites* assemblage (Figs. 47; 48). All the palynological assemblage indicate the Middle Jurassic age. The main taxa of the spores and pollen are as follows:

### Spores

1. *Neoraistrickia gristhorpensis* (Couper) Tralau, 1968
2. *Alsophilidites arcuatus* (Bolkh.) Xu et Zhang, 1980
3. *Alsophilidites* sp.
4. *Lycopodiumsporites austroclavatidites* (Cookson) Potonie, 1956
5. *L. subrotundum* (Kara-Mursa) Pocock, 1970
6. *Lycopodiumsporites* sp.
7. *Leiotriletes* cf. *adnatoides* Potonie et Kremp, 1955
8. *Deltoidospora balowensis* (Doring) Zhang, 1978
9. *Todisporites major* Couper, 1958
10. *Toroisporis* (*Divitoroisporis*) *granularis* Pu et Wu, 1985
11. *Gleicheniidites senonicus* Ross, 1949
12. *Gleicheniidites* sp.
13. *Polypodiisporites* sp.
14. *Cyathidites minor* Couper, 1953
15. *Cyathidites* cf. *concavus* (Bolkh., 1953) Dettmann, 1963
16. *Biretisporites* sp.
17. *Cibotiumspora corniger* (Bolkh.) Zhang W. P., 1984
18. *Verrucosisporites granatus* (Bolkh.) Gao et Zhai, 1976
19. *Hsuisporites rugatus* Zhang, 1965

### Pollen

20. *Alisporites minutisaccus* Clarke, 1965

21. *Concentrisporites pseudosulcatus* (Briche, Danze, Corsin et Laveine) Pocock, 1970

22. *Chasmatosporites minor* Nilsson, 1958

23. *Bennettiteaepollenites lucifer* (Thierg.) Potonie, 1958

24. *Cycadopites adjectus* (De Jersey) De Jersey, 1964

25. *Cycadopites carpentieri* (Delc. et Sprum.) Singh, 1964

26. *Cycadopites* sp.

27. *Inaperturopollenites* sp.

28. *Abietineaepollenites minimus* Couper, 1958

29. *Abietineaepollenites* sp.

30. *Podocarpidites minisculus* Singh, 1964

31. *Podocarpidites canadensis* Pocock, 1962

32. *Podocarpidites multisimus* (Bolkh.) Pocock, 1962

33. *Podocarpidites multicinus* (Bolkh.) Pocock, 1970

34. *Pseudopicea variabiliformis* (Mal.) Bolkh., 1956

35. *Pinuspollenites enodatus* (Bolkh.) Li, 1984

36. *Protopinus subluteus* Bolkh., 1956

37. *Monosulcites minimus* Cookson, 1947

38. *Quadraeculina limbata* Maljavkina, 1949

39. *Erlianpollis minisculus* Zhao, 1987

40. *Erlianpollis eminulus* Zhao, 1987

41. *Pseudowalchia landesii* Pocock, 1970

42. *Piceaepollenites omoriciformis* (Bolkh.) Xu et Zhang, 1980

43. *Platysaccus proximus* (Bolkh.) Song, 2000

44. *Quadraeculina canadensis* (Pocock) Zhang, 1978

45. *Rugubivesiculites* sp.

46. *Cedripites* sp.

47. *Protoconiferus* sp.

According to its composition, the Shaerhu flora generally has a warm temperate to temperate aspect, belonging to the early assemblage of the *Coniopteris-Phoenicopsis* flora of Northern China, and indicating an early Middle Jurassic age, probably Aalenian-Bajocian (Zhou, 1995). However, since the flora has more cycads among the mega-plants, and more subtropic or tropic taxa such as *Podocarpidites*, *Gleicheniidites*, and *Cyathidites* in the sporopollen, it shows some characteristics of the Early-Middle Jurassic flora of Southern China. Thus, the Shaerhu Flora exhibits mixed floristic characteristics of the Northern and Southern floras in the Middle Jurassic. These characteristics may be related to the fact that the Shaerhu area was geographically close to the boundary between the early Middle Jurassic Siberian flora and the European-Chinese flora (Vakhrameev, 1988). Thus, the discovery of the Middle Jurassic Shaerhu flora is most significant for the paleophytogeographical study of the early Middle Jurassic flora in China.

Concerning the formation of coal, the abundant remains of ginkgos, czekanowskialeans, cycads, and conifers in the Shaerhu flora, probably provided a sufficient material source. The flourishing ferns and other thermophilic and hydrophilous plants such as the cycads and ginkgos indicate that the Shaerhu flora probably grew in a warm and humid swamp or low wet-land during early Middle Jurassic times. The paleoclimate was mainly warm temperate with some seasonal changes which are indicated by the conifers and other deciduous leaf plants such as ginkgos, czekanowskialeans, etc. All these trees and shrubs of gymnosperms formed the forest and vigorously grew in the Shaerhu area and subsequently formed the extremely thick coal seams of early Middle Jurassic age. Therefore, in this book, the Shaerhu flora is called the "Middle Jurassic Shaerhu Forest".

# 3.9 "Top Ten Stars" of Mesozoic fossils in Xinjiang

To summarize the research findings made by the authors' research team in the past 20 years, and combined with the latest achievements of Chinese experts Xu X. and Wang X. L. in Xinjiang (Xu et al., 2006; Wang et al., 2014), the authors have singled out the following ten important fossil taxa as the "Top Ten Stars" for the newly discovered fossils in the Mesozoic Xinjiang since 2000, in which eight taxa are made by the authors' research team (Figs. 49; 77) .

The "Top Ten Fossil Stars" are as follows:

1. *Xinjiangotitan shanshanensis* (Shanshan, $J_3$)

2. Jurassic dinosaur tracks of Shanshan (Shanshan, $J_2$)

3. *Dsungarodon zuoi* (Liuhuanggou, $J_3$)

4. *Xinjiangchelys wusu* (Shanshan, $J_3$)

5. *Lystrosaurus* sp. (Dalongkou of Jimusar, $T_1$)

6. *Guanlong* Xu et al., 2006 (Wucaiwan, $J_3$)

7. *Hamipterus* Wang , 2014 (Hami, $K_1$)

8. *Nanzhangphyllum haojiagouense* (Haojiagou, $T_3$)

9. *Thinfeldia? mirisecta* (Haojiagou, $T_3$)

10. *Equisetites shaerhuensis* Dong (Shaerhu, $J_2$)

Among the above-mentioned "Top Ten Fossil Stars", the taxon of *Hamipterus* was first found in the Lower Cretaceous in Hami by Wang and his research group in 2014 (Wang et al., 2014), and *Guanlong* (theropod dinosaur) was first found from the Upper Jurassic Qigu Formation in Wucaiwan of Junggar Basin by Xu and his research group in 2006 (Xu et al., in 2006); all others were first found by the authors' research team between 2001 and 2021 (Sun et al., 2001b, 2004a, 2010, 2021; Maisch et al., 2004; Pfretzchner et al., 2005; Matzke et al., 2005; Martin et al., 2010b; Dong, 2012).

All the unique and important fossils shown in the "Top Ten Fossil Stars" including the animals and plants, together exhibit the richness and biodiversity of the Mesozoic biota in Xinjiang, with their marvelous and vigorous living scenes. The pictures have also helped us to understand the evolutionary history of the Mesozoic biotas in Xinjiang.

Fig. 77 Top Ten Star Fossils in Mesozoic Xinjiang

A. *Xinjiangotitan* (Shanshan, J$_3$); B. Dinosaur track group (Shanshan, J$_2$); C. *Dsungarodon zuoi* (Liuhuanggou, J$_3$); D. *Xinjiangchelys wusu*, (Shanshan, J$_3$); E. *Lystrosaurus* (Daloukou, T$_1$); F. *Guanlong wucaii* (Wucaiwan, J$_3$, after Xu et al., 2006; reconstruction drawn by Zhang Z. D. ) ; G. *Hamipterus tianshnaensis* (Hami, K$_1$, after Wang et al., 2014); H. *Nanzhangphyllum haojiagouense* (Haojiagou, T$_3$); I. *Thinfeldia? Mirisecta* (Haojiagou, T$_3$); J. *Equisetites shaerhuensis* ( Shaerhu, J$_2$) (H, I after Sun et al., 2001b, 2010; J after Dong, 2012)

# Chapter 4

# Emotion with fossils in Xinjiang

## 4.1 Affection with the fossils in Baiyanghe

The abundant fossils and their origin in the Baiyanghe (Baiyang River) area in Tuoli, Xinjiang, are just like the "fragrant grass" falling from the sky, bringing endless joy and nostalgia to the experts of the Sino-German cooperative scientific research team.

The magic of the Baiyang River lies first in the fact that it looks like an oasis in the vast Gobi desert. From Tiechanggou in the west of the Baiyang River to Hestuologai in the east, on the vast Gobi stall spanning about 200 km, it is the only green spot that can be found. The patches of Baiyanghe forest make people feel the tenacity of life. The wonder of the Baiyang River valley lies in its rich and wonderful fossils: they record the story of life in the Junggar Basin of Xinjiang more than 100 million years ago, and also provide a precious key to uncover the "Junggar mystery" of geologist Obruchev. In addition, over the past 20 years, Baiyanghe has been left with solid footprints and hard sweat of the members of the Sino-German research team, as well as many unforgettable stories, which make us feel endless attachment.

## 4.1.1 Entering Baiyang River seven times

Baiyang River in Tuoli County is located about 50 km east of Tiechanggou of Emin County. It originates from the high Mt. Shaleketeng and flows all the way down to the south, and after passing through the Baiyang River valley ("Ak-djar ravine"), the water flow becomes gentle in the Gobi. In summer, there is less rain, so one can easily cross the river. However, when the snow melts in spring or the rainy season sets  in autumn, the water here will be knee-deep, and it becomes a little difficult to cross the swift river. Since the main geological section yielding the fossils is on the east bank of the river, the members of the research team have to cross the cold river every day from their camp which was on the west bank. It is difficult to wade across the cold river every morning, and there were still the rolling gravels at the bottom of the river. Sometimes it is not easy to cross the river with a wooden stick.

In order to collect more fossils and further unravel "Obruchev's secret", Chinese members of the research team headed by Sun G. have visited the site seven times since 1989. When crossing the river in the morning, the researchers were loaded with all kinds of field equipment, including steel drill, hammer, measuring rope, and cartons full of wrapping paper. While at dusk, when field work ended, the boxes with heavy fossils and rock specimens had to be brought back. Sometimes the fast-flowing river rose over the knee, and the feet were prone to slip due to the moving gravel at the bottom of the river at any time. In spring and autumn, the cold water not only poured into the boots but also soaked the trousers. In October 2015, when the research team came to Baiyang River for the sixth time, due to the large number of fossils collected, the weight of each box was more than 30 kg. Although the youngest Dr. Liang F. in the team was over weight, he carried the heaviest fossil specimen boxes on his shoulders. Nevertheless, after crossing the river he had to return to help his supervisor Sun G. and his colleagues. There's an old Chinese saying that "the youngest people should suffer more". Since Liang F. was the youngest in the team, he always took on harder work and kept enthusiasm as his pleasure in the field work (Figs. 50; 78).

In early May 2018, several members of the research team took the holidays of the "Labour Festival" for the seventh field trip to the Baiyang River. Due to the melting snow, the Baiyang River was waist-deep, and it was impossible to cross the river. For this reason, they changed their plan and moved to Hestologai Town for "camping", which was about 200 km east of the Baiyang River. Thus, every day they spent 6 or 7 hours in the Gobi "way" to and back from the Baiyang River fossil site by car, though there was no real road in the vast Gobi. To ensure the working hours, they got up early and returned late. When they finished their work and returned to the station, it was already nightfall every day. But the researchers never felt daunted in face of all these difficulties, especially the young men. This is because they mainly wanted to cherish their precious time, collect more fossils, and strive for new discoveries.

## 4.1.2 "The toothache road"

The distance from Tiechanggou to the Baiyanghe fossil site is about 50 km in eastern direction, including a 30 km asphalt road in the beginning, and then followed by 20 km gravel roads in the deserted Gobi. On the uneven gravel road the car was jolting, and the gravel hit the body of the car.

In October 2015, when the research team came to Baiyanghe for field work, it so happened that Prof. Sun G., the co-leader of the team, was suffering from serious dental problems, intermittently with unbearable pain. Because Tiechanggou was a small and remote town without senior dentists, it was necessary to drive more than 100 km to the hospital in Karamay City in order to find an experienced dentist. To save valuable working time in the field, Sun G. had to stay in the torments of pain, still working in the field. Whenever the car came down from the road and entered the Gobi's gravel road, which was about 10 km from the Baiyang River section, the bumps on the road made Sun's toothache worse; sometimes the car had to pull over and wait for Sun to take analgesic tablets before it continued its journey. However, miraculously Sun's toothache was often relieved immediately when the car left the bumpy gravel road and entered a smooth asphalt road. Thus, this section of the gravel road on the Gobi was called in a funny way "toothache road" by the team members,

Fig. 78 Entering the Baiyang River valley for seven times

A, I. Baiyang River valley and its geological section; B. Crossing the river; C-F. Dr. Miao Y. Y. and her group working in the section in 2005, E showing her collected *Ginkgo* fossils in field; G, H. Prof. Sun G. and his group working in the section in 2015; J. Sheeps in Baiyang River valley (2005)

and it was hoped to have a signboard erected at the starting point of this section of the road, on which there are these words: "driving through the toothache road, the road ahead is a thoroughfare".

### 4.1.3 Training of graduates

Dr. Miao Y. Y. was a graduate student for MSc. and Ph. D. in Jilin University of China and she had Prof. Sun G. as her instructor from 2001 to 2006. Her Master and Ph. D. theses both focused on the study of the Middle Jurassic Baiyanghe flora.

In May 2005, Miao came to the Baiyanghe fossil site to carry out the field work for her doctoral dissertation, accompanied by three colleagues. In early May, it is still quite cool in the mornings and evenings in the Junggar Gobi. In the morning, when crossing the Baiyang River, she wore a winter jacket and waded the cold river with bare feet and rolled-up trousers. By noon, the sun was scorching hot overhead, the rocks on the outcrop were burning, and accidentally small stones fell from the section, so the working conditions were quite tough. But these difficulties were nothing to her. Miao was born in Harbin of Heilongjiang Province, in northeastern China. After graduation from China University of Geosciences in Wuhan, she worked in the Geological Survey of Heilongjiang for several years and withstood the harsh environment in the wild. However, in the Heilongjiang region, there are dense forests under a humid climate. The dry heat and sandstorms in the Gobi of Junggar were a quite different challenge, and became a severe test for her. After a month of field work at the Baiyanghe fossil site, her face was tanned and her hands got blistered. However, she took the hard work as a good opportunity to temper her will. Fortunately, her scientific research work achieved fruitful results, and she learned also to ride a horse from the local Kazakh herdsmen. After the completion of field work in Baiyanghe, she successfully accomplished her indoor research and thesis writing, and received her Ph. D. degree in 2006. Since then she has published several papers on the Middle

Jurassic flora of Baiyanghe. Whenever she recalled the field work in the Baiyang River valley, she was filled with emotion that the field work in Baiyanghe provided unforgettable training, and over the years, she has had a strong attachment to the fossils of Baiyanghe ( Figs. 50; 78).

## 4.2  Entering the Gobi of Shanshan three times

The climax of the cooperative research work of the Sino-German scientific research team was the "three visits to the Shanshan Gobi" between 2007-2013: searching for dinosaurs, turtles and other fossils in Shanshan of the Tu-Ha Basin. The Shanshan Gobi has witnessed the valuable spirit of the members of the Sino-German research team in their painstaking and persistent search for fossils, and also left behind many moving stories.

### 4.2.1  First visit to the Gobi of Shanshan

Since 2007, the key work of the Sino-German research team was to search for dinosaur fossils in the Tu-Ha Basin. The initiative for this came from the 2nd meeting of the Sino-German joint laboratory held in Frankfurt, Germany in November of 2006. At the meeting, Prof. Zuo X. Y., the Ex-Director of RGSX No. 1, suggested looking in the Turpan Basin of Xinjiang for dinosaur fossils based on his work experience in this area. In early September 2007, the research team held the "International Symposium on Geology and Environmental Evolution of Northern China" in Urumqi, with a post-symposium field trip to Turpan, which provided an opportunity to carry out this proposal. Thus, after the symposium, all of the members of the research team visited the Turpan area for the first time, and a small group stayed to work in some detail in this area, led by Dr. Wings O., a German young expert, and Prof. Dong Z. M., the famous dinosaur expert of China. The group consisted a total of six members, five of whom were young people from Germany who had looked forward to field work in the Tu-Ha Basin for a long time. The group's first stop was Shanshan County, about 90 km east of Turpan City. This area was the place where Prof. Dong discovered several dinosaur fossils in the period 1970s-1990s (Fig. 51).

Prof. Dong graduated from Fudan University in Shanghai and entered the IVPP in 1963 as a graduate under the supervision of Academician Prof. Yang Z. J. (Young C. C.). For more than 50 years, Dong has traveled all over China, mostly in Inner Mongolia, Xinjiang, and other remote areas of China, and has discovered many dinosaur fossils, particularly in 1970s-1990s, including the theropod *Shanshanosaurus huoyanshanensis* Dong, 1977 from the Upper Cretaceous, and the sauropod *Hudiesaurus sinojapanorum* Dong, 1997 from the Upper Jurassic, both in the Turpan area (Dong, 1993, 2001, 2009). This time, when he led the group to Shanshan again, it looked like his "return to hometown", and he cherished the wish to make new discoveries. Dr. Wings O., the leader of the young scientists from Germany, was a young expert in dinosaur research. He graduated from the University of Tuebingen and worked in the Natural History Museum in Berlin at that time, and was energetic with the courage to forge ahead. Other people from Germany include Schellhorn R., Koerner K. from the University of Bonn, and the technician Mueller M. Although they came to the Shanshan Gobi of Turpan for the first time, they were full of enthusiasm to find dinosaur fossils. What's more, they brought a lot of fossil mining experience from Germany.

In September, Shanshan Gobi is under the scorching sun. To its south, the vast and famous Kumutag Desert, one of the eight deserts in China, stretches about 400 km. The ground in the hot Gobi and desert seems to be baking in a furnace, and the temperature can sometimes reach nearly 50 ℃ . Prof. Dong guided the group to Qiketai, about 20 km east of Shanshan Town, where the Gobi stretches for tens of kilometers, formed by purple red rocks intercalated with grayish green rocks spreading out in stripes, and wide areas covered by dark gravels. The group traveled along the north-south ravines to find the dinosaur bones discovered by Prof. Dong more than 20 years ago, and it was just like looking for a needle in a haystack. But in any case, Prof. Dong pointed out the direction in the field work in the Shanshan Gobi (Fig. 79, B).

In this way, after Prof. Dong left Shanshan, Wings and his partners, accompanied by Mr. Du Q., a geologist from the RGSX No. 1, worked tenaciously on the Gobi for several days, and finally, they made a new discovery in a narrow gully south of Qiketai, Shanshan—the discovery of the first Jurassic dinosaur tracks in the Turpan Basin.

## 4.2.2  Unexpected lunch

When Prof. Dong Shanshan, the German young men were reluctant to part with him. The next morning, Wings and his partners set foot on the Gobi south of Qiketai of Shanshan again. They crossed several north-south gullies composed of variegated and red rocks, and climbed up and down dozens of times until they were somewhat exhausted. Toward noon, they decided to rest and have lunch at a slightly open gully (Fig. 52, A).

Several young people went up the east slope of the ditch, sat down on the earth slope, and began to eat, while a German young man suddenly noticed that the yellow-green rock outcrops on the opposite cliff appeared colorful in the sun, much like a rock mural. Thus, he put down his lunch box and hurried to climb up to see. How surprised he was when he realized that the "rock mural" on the yellowish-greenish sandstone cliff exposed dinosaur tracks! He immediately shouted for everyone to see, and all partners looked at this magic discovery. After taking a close look, Wings confirmed that these are real dinosaur tracks densely crowded and highlighted by the sun on the yellow-green rocks. Thus everyone was suddenly very excited and cheerful at this new discovery! Soon after, Wings asked everyone to pursue north along the direction of the track-bearing beds, and as expected, more dinosaur track fossils were found on a larger rock outcrop extending about 30 m in the gully; and after a preliminary calculation, more than 130 dinosaur tracks were found in total! Thus, the largest known Jurassic dinosaur track site in China was accidentally found, and this discovery was in the middle of the unexpected lunch in the Shanshan Gobi ( Figs. 52; 79, C-G)!

Soon after, Dr. Wings reported the news of this important fossil discovery to Prof. Sun G., the co-leader of the research team. Sun G. with Dr. Wu hurried to Qiketai of Shanshan to look at the newly discovered dinosaur tracks and very much appreciated this new finding, and immediately reported this discovery to the local government. Later, Prof. Dong, Prof. Martin, and other experts also rushed to the scene and confirmed that this was the first discovery of Jurassic dinosaur tracks in the Mesozoic of Xinjiang, even in China. After that, the follow-up research on this dinosaur track group was also carried out intensively, including careful description and initial classification. The paper was published in the journal *Global Geology* in the autumn of 2007, led by Wings, and joined by Prof. Dong Z. M., Dr. Wu W. H., and other participants of the research team.

On April 8, 2008, initiated by the Sino-German Research Team, the Shanshan County Government held a press conference in Shanshan, officially announcing the first discovery of the largest known Jurassic dinosaur track group in Shanshan. Prof. Sun G. and Prof. Martin, together introduced the new discovery scientifically at the press conference. Prof. Li J. J., the Chinese expert on dinosaur track research from the Beijing Natural Museum, delivered a speech to commend for this discovery. The media such as Xinhua News Agency and the local major media gathered at the conference and quickly reported the news of this new discovery. The first discovery of a Jurassic dinosaur track assemblage in Shanshan not only fascilitated the study of

Fig. 79  The new discoveries of vertebrate fossils in Shanshan of Tu-Ha Basin

A. The Sino-German Research Team entering the Shanshan Gobi (2007); B. Prof. Dong Z. M. guiding Dr. Wings in the field work in Shanshan; C-G. Jurassic dinosaur track group found in Shanshan (2007); H. The largest known Jurassic dinosaur *Xinjiangotitan* found in Shanshan (2009- 2013); I. The Jurassic turtles of *Xinjiangchyles wusu* found in Shanshan (2013)

the Jurassic dinosaur fauna and its paleoecology in China, but also filled a gap in the research of dinosaur tracks in Xinjiang (Fig. 53).

### 4.2.3 Meeting with turtle *Xinjiangchelys*

The second phase from 2009 to 2012 to enter the Shanshan Gobi was mainly intended to find turtle fossils and expand the search for dinosaurs. In early April 2009, German young experts Wings O. and Joyce W. led young scientists and students Schellhorn R. (UT), Frank N., Lenssen L., Koch M., Schwermann A., and Hoffmann S. (UB), accompanied by Sun G. and Yang T. to the Gobi in Qiketai of Shanshan, again. After hard work, the German young experts found several turtle fossil sites on the top of a small hill (42°57'31.5"N, 90°33'21.0"E) about 10 km south of the dinosaur track site.

To make further achievements, in 2010 and 2011, when the young expert Joyce W. was granted new research funds, with the support from and arrangement by Prof. Sun G., the German young group headed by Joyce and Wings. came to Shanshan again. The group also included the graduates Rabi M. (Hungarian) and Regent V. (as a technician), and Dr. Zhou C., the young Chinese expert, and Dong M., a graduate student supervised by Prof. Sun G. (Figs. 28; 29). The leaders and colleagues of the Shanshan County government took warm care and provided generous support. The Deputy county director Mrs. Zhou herself created favorable working conditions for the young research group, vacating two rooms from the local storage for the storage of fossils, and arranging a site to provide convenience for the group to prepare fossils. She also accompanied Sun G. and the young experts to the field site in Qiketai (Figs. 54, H; 59, H).

During field work in October 2011, the young group consisting of Wings, Zhou, and Rabi among others discovered seven new fossil turtle sites in an area of 10 km$^2$ in the Gobi of Qiketai. In October, the baking sun was still intense in the Gobi of Qiketai, and a mixture of wind, sand, and rock dust pervaded the area, which made it sometimes difficult to open eyes. Besides, the turtle-bearing rocks are very hard, which made collecting turtle fossils quite difficult. To completely open the face of the block containing the turtle fossils, two female graduates worked hard in the lab for preparation. German Regent V. and Chinese Dong M. together carefully prepared the fossils, and they finally found more than ten well-preserved turtle shells in one block (Figs. 28; 73). In addition, they also found some dinosaur partial skeletons and more than 20 new fossil sites in Qiketai Gobi. Learning about the exciting news from Zhou, Wings and Joyce, respectively, Prof. Sun G. was very enthusiastic and began planning a larger-scale excavation of dinosaur fossils in this area, which was the prelude to the large-scale dinosaur fossil excavation and the "three enterings Shanshan Gobi" carried out by the Sino-German scientific research team in the Tu-Ha Basin.

It should be mentioned that due to the hardness of the fossil-bearing rock, it sometimes took 1-2 months to prepare an intact fossil. As shown in figure 54 D, the completely preserved turtle was prepared by German graduate Regent V. in the PMOL in Shenyang, China, for a whole month. During her work, the white dust of rock lime covered her face, and the beautiful face of Regent often seemed to be painted with white powder, just like the "clown" actor in the Chinese Peking Opera. Thus, after hard work, the specimens of *Xinjiangochelys wusu* are presented beautifully (Fig. 54).

# 4.3 "Voice of Dinosaurs" in Turpan

## 4.3.1 The "voice of dinosaurs"

As the "Second Entering Shanshan Gobi" got the clue of finding dinosaur skeletons in the Gobi of Qiketai in Shanshan, the Sino-German research team has carried out large-scale dinosaur excavations in the Gobi of Shanshan since 2012. These operations were strongly supported by the National Committee of Fossil Experts (Beijing), the Department of Natural Resources of Xinjiang (DNRX), and the Turpan

Region and Shanshan County governments, and were carried out with the official approval of the Ministry of Natural Resources, China (MNRC). As the year 2012 was traditional "Dragon Year" in China, the research team named the excavation the "Voice of Dinosaurs in Xinjiang", to hear the call of the dinosaurs as soon as possible and let the voice of the dinosaurs of Shanshan sound all over Xinjiang, China and even the world.

The large-scale dinosaur excavations in Shanshan began in the spring of 2012 and continued until 2013. More than 20 people participated each time, and Prof. Sun G. was personally leading the excavation in the field. The Shanshan Gobi stretches for nearly 100 km. To the south, it is adjacent to the Kushtag Desert, one of the eight largest deserts in China. We experienced dry and also rainy conditions. In April, the temperature was still very low at night, but the earth was roasted at noon, which was unbearable. To improve the working efficiency, the team members prepared breakfast and lunch in advance the night before and got up at dawn, so that we could work more efficiently in the morning before the heat of the day.

The size of the large-scale dinosaur excavation in the Gobi south of the Qiketai was more than 1000 m$^2$. The dinosaur-bearing rocks were hard, mainly due to calcium-carbonate cementation of the rocks, which made the excavation difficult, but which did not diminish the enthusiasm of the team. During the work, the young people of the research team used pickaxes and shovels to clean up the surrounding earth and rock, and sometimes even used an electric rock drill. When dinosaur bones came to light, they used chisels, small shovels, and brushes to do "fine work" in excavation. It was most touching to see the famous expert Prof. Dong, at an age of over seventy years, running up and down the hill to guide the excavation. A young team member Dr. Sekiya T. from the Fukui Dinosaur Museum of Japan joined the excavation. He braved the scorching sun, took great pains to measure and record the exposed dinosaur bones, and completed several exquisite sketches and restoration drawings of the fossils. The

German young men also worked very hard, and were showing a beautiful "scenic spot" when they took off their shirts during work in the hot sun and were covered by the wind-blown dust which turned them into "pink people" (Fig. 55).

In this way, first the femur, pelvic girdle, sacral vertebrae, and other bones were lifted, and then the dorsal and cervical vertebrae were also revealed. In particular, one of the large femurs measuring about 1.6 m in length was quite spectacular! According to Prof. Dong's estimate, the newly discovered large dinosaur should be about 30 m long and would be the largest and the best preserved Jurassic dinosaur in China. Looking at the discoveries that came to light day by day, all the members were very excited!

Under the guidance of Prof. Dong, through cooperative and careful research, the new discovery was confirmed scientifically, and the research team members named the dinosaur as *Xinjiangotitan shanshanensis*, which is the best gift given by the Sino-German research team to the host places Xinjiang and Shanshan, respectively. In September 2013, the discovery results were officially published in the journal *Global Geology* with the authorship of Wu, Wings, Zhou, Sekiya, and Prof. Dong. The follow-up excavations of this dinosaur were made by Prof. Li D. Q., the Chinese expert on dinosaur research, and his group, with significant progress by the discovery of the incomplete skull, additional 15 cervical vertebrae (18 in total with the previous findings by the authors' team) and 39 caudal vertebrae, which further improved the research level for this giant dinosaur (Zhang et al., 2018). The new discovery of *Xinjiangotitan shanshanensis* represents the largest known Jurassic dinosaur in China, creating a new record of Jurassic large dinosaurs in China. It is of great significance to further deepen the knowledge of the evolution of dinosaurs in China and the Jurassic paleogeography and paleoclimate in Xinjiang, and also for the fossil protection in Shanshan. Not long after, the Jurassic Museum of Shanshan and the National Geopark have both been established in Shanshan County, and all the members of the Sino-German research team seem to have heard the earth-shaking roar of the giant dinosaur in Xinjiang.

## 4.3.2 Festival in Shanshan

To congratulate on the discovery of the largest known Jurassic dinosaur in Xinjiang, even in China, and to promote the development of fossil protection in Shanshan and Xinjiang, on October 10, 2013, the Shanshan County Government held the grand "Ceremony of the naming of *Xinjiangotitan* and the meeting for fossil protection" in Shanshan. More than 100 people participated in this ceremony and meeting, including Chinese officials, members of the Sino-German research team, and guests from the University of Bonn, headed by Prof. Martin T. and Prof. Litt T. (Fig. 56). During the ceremony, people in Shanshan presented flowers to the meritorious experts who found the dinosaur fossils in the Shanshan Gobi, and the local governmental leaders delivered enthusiastic speeches and warmly expressed their thanks to the Sino-German research team for their hard and successful work. Colorful flags and balloons were flying in the air at the venue, and Shanshan celebrated an exciting festival.

Finally, the "Emotion with fossils in Xinjiang" has borne sweet fruits. Whenever they recall those unforgettable days excavating fossils in Xinjiang, the Sino-German research team members are always filled with nostalgia and incomparable pride.

# Chapter 5

# Sino-German friendship

## 5.1 "The second life"

The most unforgettable story of the Sino-German cooperative scientific research team in Xinjiang is the "second life" of Prof. Ashraf A. R. from Germany, one of the key members of the research team. He was accidentally injured and seriously fractured his hipbone during field work in Xinjiang. The story of his rescue operation in China and full recovery has been widely told in China and Germany. He often said, "Xinjiang gave me a second life!"

Rahman Ashraf is a German palynologist and a very close relative of the latest Afghan M. King Zahir Shah "Father of the nation". He graduated from the Dept. Geology of Kabul University in 1966 and later moved to Germany in 1967, and received his Diploma "Dipl. Geol." in 1972 and Ph. D. from the University of Bonn in 1977, under the supervision of Prof. Wurster P. and Prof. Schweitzer H.-J., the famous paleobotanist. Since 1978, he has been successively engaged in scientific research at the Institute of Paleontology of the University of Bonn and later in the Institute of Geosciences of the University of Tuebingen for more than 20 years. He was a Visiting Professor of the Dieleman University of Manila, Philippines from 1989 to 1991, and has been an Honorable Professor of Jilin University, China since 2001, and Honorable Professor of Shenyang Normal University, China since 2011. He was Co-Vice-Director of the Sino-German Geological Co-Working Station in Xinjiang (SGWSGX) from 2000 to 2009. Over the years, he has made important contributions to palynology and biostratigraphy both in Germany and China, particularly to the studies of the Mesozoic palynoflora and strata of Xinjiang, and the K-Pg (K-T) boundary in Nanxiong of Guangdong and Jiayin of Heilongjiang, China. He has visited China many times and played an important role in the Sino-German scientific cooperation and cherishes profound feelings for China and the Chinese people. His excellent character combines in a unique way the traditions and cultures of the East and the West, and is respected and praised by all the Sino-German research team members (Figs. 61; 62).

One day in July of 1997, he was guiding a group during field work in an old pithead closed for many years near Xishanyao about 50 km west of Urumqi. When he explained the geology of this place to a young German graduate, he touched the rock on the top of the pithead with his hammer. All of a sudden, the pithead wooden pillar collapsed, and heavy rocks fell down, injuring seriously his hip bone. The German graduate immediately called for help, and all the colleagues came for rescue immediately. However, the huge rocks were too big and heavy, and could not be moved by several people. Finally, the nearby coal miners helped to move the rocks and everyone struggled to carry him to a car with a self-made stretcher for secure transport. He was then immediately driven to the hospital in Urumqi for treatment. The distance from the Xishanyao area to Urumqi was only about 20 km across the Gobi, to reduce the harm of jolting on the severely injured, the car traveled slowly for nearly 4 hours! It was late at night when the car arrived in Urumqi, at the Army

General Hospital, the best hospital in Xinjiang.

As Prof. Sun G. was ill and could not go to the field, he was anxiously waiting on the steps at the gate of the guest house, expecting everyone to come back from field work in the early night. When he learned about Prof. Ashraf's injury, he hurried to the hospital. First, he went to the Dept. of Radiology of the Emergency Unit of the hospital to see the X-ray film which confirmed that Ashraf had suffered a serious hip fracture. Then he immediately rushed to the ward to see Ashraf, and when Sun Ge met Ashraf, all the colleagues present shed tears, because Sun G. and Ashraf were very close friends for many years. That night, Sun G. immediately called the NIGPAS for extra help, and the next day, young Dr. Wang Y. flew from Nanjing and took turns with Dr. Wang X. F., a Ph. D. graduate of Prof. Sun G. and a member of the research team to look after Prof. Ashraf. The doctors of the hospital worked very hard, and the X-ray equipment used was the most advanced, made by the Siemens Company of Germany. It became evident that Ashraf's right hip fracture-dislocation was about 2 cm, which was a serious case. To get timely and effective treatment for Ashraf, the Army General Hospital made available a single ward for him for best treatment and full recovery. As advanced communication equipment was not common in China at that time, Prof. Zuo brought a mobile phone for Ashraf for long-distance calls to his mother in Switzerland. When Ashraf was talking to his mother, he and every colleague present had tears in their eyes. During the ten days of Ashraf's hospitalization, besides the members of the research team, the leaders, and colleagues of the RGSX No. 1 came to visit, and presented flowers to Ashraf. The ward in the hospital was turned into a caring family. Ashraf was deeply moved by the warm care and friendship in the hospital in Xinjiang, China.

To ensure a quick recovery of Ashraf's bone injury, with the efforts of Prof. Mosbrugger V. , co-leader of the Sino-German research team, the German insurance company ADAC sent a special plane with four staff members (2 doctors and 2 pilots) to Urumqi, China, to pick up Ashraf and to return to Germany for further treatment. Sun G. and Mosbrugger V. rushed to prepare the information for the German insurance companies in China and Germany respectively. On July 19, when Ashraf's stretcher was placed next to the apron and the friends and colleagues said goodbye to him, tears filled everyone's eyes. Finally, when everyone learned that Ashraf had safely returned to Germany and that his fracture had healed after careful treatment in the Hospital of the University of Tuebingen, all Chinese colleagues felt very relieved and cheerful for Ashraf.

Since the field accident occurred on July 9, and Ashraf's birthday is on July 2, Ashraf has celebrated July 9 as his "second birthday" since then. Therefore, whenever he came to China later, he always said that Xinjiang of China gave him "a second life". In December 2000, when Ashraf returned to Xinjiang after his recovery, Prof. Sun G. presented a calligraphy banner written by himself to Ashraf at the welcome meeting in Urumqi, saying *"Xinjiang has a long summer and brotherhood"* in Chinese (Fig. 57).

## 5.2 "Sino-German family"

"When we come to Urumqi, we return to our second home!" These are emotional words that the Sino-German research team members from Germany say every time they return to Xinjiang. Indeed, thanks to the dedicated efforts of the leaders and colleagues of the Xinjiang geological and mineral system, the members of the SGWSGX were becoming a warm family of scientists from China and Germany. The "big family", headquartered in the RGSX No. 1 in Urumqi, did a lot of work for the geoscientific cooperation between China and Germany for over 20 years, so that the scientists from China and Germany did not only cooperate closely academically but also established a deep friendship (Figs. 58; 59).

To arrange the work of the SGWSGX station, the host, RGSX No.1 specially vacated a separate building as the official place of the working station, and arranged for Li J., a local paleontologist, and Guo H., the ex-head of the office, responsible for the reception. To facilitate the field work, the RGSX No.1 prepared several off-road vehicles, which were the best and safest models in China at that time. The first stop of the Sino-German research team was the Haojiagou, about 50 km from downtown Urumqi, and half the way was spent on asphalt roads, and the other half on rural dirt roads. Sometimes in places where cars could hardly move, people either changed to carts or debarked and walked without complaint. The Haojiagou is located in a remote valley, where only the Kazakh minority herdsmen Hadley and his brother live. Sometimes the research team collected many specimens and rock samples from the mountains, which were usually too heavy to carry. The Hadley brothers always came to help with their horses, and in return, the research team presented some gifts to them. Gradually, they have become old friends.

The SGWSGX operated in Xinjiang for about ten years (2000-2009), and German experts (including students coming for geological practice) came to Xinjiang more than ten times. During this period, Chinese experts (including the staff of the RGSX No. 1) were also invited to visit Germany for many times, including Universities of Tuebingen, Bonn and Frankfurt, and the Senckenberg Natural History Museum. To cultivate talents, under the friendly arrangement of Prof. Mosbrugger, the Chinese young staff member Guo H. was invited to take a short-term study at the University of Tuebingen in Germany for a month, which was funded by the German side. This study in Germany expanded Guo H.'s international vision and knowledge, so that she was later recruited by the Office 305 of Xinjiang in recognition of her growing talent. The exchange of visits between Chinese and German counterparts has played an important role in improving the scientific level and promoting international exchanges of the RGSX No. 1.

## 5.3  Growth of young people

It is often said that the cactus (*Opuntia dillenii*) is a "desert flower", and *Populus euphratica* is the "desert God". Cactus likes light, bears drought and is easy to grow. Its fruit is purplish red with many seeds. It is pleasing to the eye when its flowers bloom. *Populus euphratica* has deep roots and luxuriant leaves in the harsh desert, and people admire them for their tenacity in their life. The growth of young people from the Sino-German scientific research team in Xinjiang is like these two plants.

During the period when the Sino-German research team was working on the Mesozoic of Xinjiang, a number of excellent young scientists have thrived, such as Miao Y. Y., Liu R., Wu W. H., Xu Y., Yang T., Dong M., Bian W. H., Zhang J. G., Meng Q. T., and Zhou C. F. from China, Sekiya T. from Japan, who was a graduate of JU, China also, and Maisch M., Matzke A., Wings O., Joyce W., and Frank V. from Germany, and Rabi M. from Hungary, who was also a graduate of UT, Germany. They were all trained or supported by the Sino-German cooperative research work in China. Now, they have become professors, senior scientists with Ph. D. or MSc. and have achieved recognition in research, teaching, museums, or other working positions. At the same time, these young people have also contributed to the study of the Mesozoic of Xinjiang, and the development of the Sino-German friendship (Fig. 60).

It should be mentioned that thanks to the cooperative work of the Sino-German research team in Xinjiang, several Chinese young people have had the good chance to go to Germany for further advanced study, which has enabled them to broaden their professional knowledge and learn new methods and techniques in research from the German senior scientists. For example, Bian W. H. and Zhang J. G. both visited Darmstadt University of Technology, Germany, and received very kind instructions from Prof. Hinder M. and Dr. Hornung J. to promote their scientific qualification (Fig. 60, C and I). Dr. Meng Q. T. had enthusiastic guidance and help in palynological research from Dr. Bruch A., and then she much improved her research level in paleoenvironment oil shale from Northeast China, with new achievements. Besides, Drs. Zhou C. F. and Wu W. H. have also got more help in paleovertebrate research from Prof. Martin both in Germany and China.

## 5.4 Honors for the scientists from both sides

The development of the Sino-German geoscientific cooperation has promoted the scientific research and teaching work of China and Germany. Over the past 20 years, Chinese and German scientists have received honors in each other's countries. The German academician Prof. Mosbrugger and Prof. Ashraf have been awarded Honorary Professors of Jilin University and Shenyang Normal University of China; Prof. Martin has been awarded a Guest Professor of Jilin University and Shenyang Normal University. Named after Prof. Ashraf, the RCPS of Jilin University and the PMOL of Shenyang Normal University have set up the "Ashraf Laboratory" respectively in honor of his donation of thousands of reprints and books, as well as some research equipment to the two institutions. In addition, Prof. Hinderer and Dr. Hornung were awarded Guest Professors by Jilin University of China (Figs. 61; 62).

On the German side, on July 3, 2018, the Senckenberg Research Institute and Natural History Museum held a ceremony in Frankfurt Squaire, on behalf of the Senckenberg Society for Nature Research Germany (SGN), and conferred the honorary title of being a Corresponding Member of the Society on Prof. Sun G. Prof. Sun G. is the first Chinese recipient of this honor (Fig. 61, G). Besides, Prof. Sun G. also received an honorary certificate of the Institute of Geosciences, University of Bonn, awarded by Prof. Martin (Fig. 61, H). The Senckenberg Society of Natural Research was founded in 1817 and is one of the most famous academic societies in the world. In the past, the honor of Corresponding Member was awarded to famous scientists and philosophers such as von Goethe J. W., Darwin C., Cuvier G., Hegel G., and Humboldt A., etc. The award of this honor to Prof. Sun G. shows that the SNG attaches great importance to the Sino-German cooperation and it is also a symbol of the Sino-German friendship.

In addition, in October 2010, the Sino-German Science Promotion Center held a big celebration for the 10th anniversary of its establishment in Beijing, China. Prof. Sun G. and Prof. Mosbrugger V.were specially invited to jointly present the only working report in this conference, which reflects the high attention and courtesy of China toward the Sino-German cooperative research team headed by Prof. Sun G. and Prof. Mosbrugger V. (Fig. 62, B).

## 5.5 Flowers of Sino-German friendship

The geoscience cooperation between China and Germany in Xinjiang has been expanding and attracting the participation of scientists from the United States, Japan, Austria, Hungary, Luxembourg, and Ireland. The achievements of the Sino-German cooperative research on the Mesozoic of Xinjiang, China have spread far and wide and been quoted internationally.

The achievements and friendship demonstrated in the Sino-German cooperation in Xinjiang are colorful flowers blooming to the world with their tenacious vitality and beauty. In the past 20 years, the Sino-German scientific research team has received enthusiastic support and participation from Prof. Dilcher D. L. (NAS, US), Prof. Nishida H. (Japan), and scientists from 13 institutions of China and Germany (Figs. 8; 9; 63, B and C).

Besides, the Sino-German cooperation in Xinjiang has also built a bridge between universities and museums in China and Germany. In 2014, a grand "Special Exhibition on Feathered Dinosaurs from China" was held at the State Natural History Museum in Stuttgart (SNHM), Germany in cooperation with the Paleontological Museum of Liaoning (PMOL), China (Fig. 63, E and F). Leaders of both museums are members of the Sino-German research team (i.e., Sun G. as the Director of PMOL, and Prof. Eder J. as the Director of SNHM) (Fig. 80, J). In the summer of 2017, 26 teachers and students from University of Bonn, Germany visited China for field geological practice. They not only went on a geological field trip to western Liaoning

Fig. 80  Flowers of Sino-German friendship blooming

A. Sun G. and Mosbrugger V. visiting Köln, Germany (1988); B. Field work in Changbai, China (2006); C. Working in Urumqi, China (2004); D. Field work in Fushun, coal-mine of China (from left): Eder J., Sun G., Mosbrugger V., Bruch A. (2006); E. Ashraf in Urumqi. Xinjiang (1997); F. Instructing Chinese graduate in Jiayin, China (2008); G. Chinese Ph. D. graduate Bian W. H. with his supervisors Prof. Hinderer (right) and Dr. Hornung in field work; H. Prof. Thein J. guiding field work in Germany (2006); I. Prof. Sun Y. W. giving a talk in the symposium in Urumqi (2004); J. Prof. Eder J. (left 3) hosting Chinese guests visiting Stuttgart Museum (NH) (2013); K. Visiting Xinjiang Geological Museum, guided by Prof. Song S. S. (ex-director of RGSX No. 1), joined by Prof. Dilcher D. L. (NAS, second from right) et al. (Urumqi, 2004 )

and Jilin, but also collected fossils in Jiayin of Heilongjiang, China, and attended China's academic conference. The visit was very successful and highly estimated by the German teachers and students and strengthened the friendship between German and Chinese students and teachers (Fig. 63, G).

In addition, Chinese and German scientists have also enhanced their friendship and made friends with the local people in Xinjiang, especially the ethnic minorities near the fossil-yielding places. In September 2009, Profs. Mosbrugger V. and Sun G. were invited to visit a local Uyghur family in Shanshan where the giant dinosaurs were found, and enjoyed the hostess' warm hospitality, including eating the famous Turpan grape and Hami melon (Fig. 58, D). In October 2013, the local minority Tajik people in the Dalongkou village (the P/T boundary point) of Jimsar warmly welcomed the experts from University of Bonn, Germany, and they took a group photo together with the boys and girls happily joining also. The precious photos left a nice memory for the local people and Chinese and German experts (Figs. 63, H and I).

The Sino-German cooperation in Xinjiang and its scientific achievements were like a ribbon connecting the cooperation in Geoscience research between China and Germany, and the Central Asian countries, Russia, Mongolia, and other countries. Not long ago, the Institute of Geology of the Kyrgyz Academy of Sciences proposed cooperation on the comparative study of dinosaur fossils between Kyrgyzstan and Xinjiang, China, and invited some experts of the Sino-German research team. All these developments indicate that the international cooperative research on the Mesozoic of Xinjiang, China will be further carried on in the future.

May flowers bearing Sino-Germany friendship always bloom!

# Acknowledgement

The authors would like to express sincere thanks to the support for the Program of CAS (China)-Max Planck Society (Germany) from 1997 to 1998; the Major Int'l Cooperation Project of NSFC (30220130698) and other projects of NSFC (China) and DFG (Germany) since 1999; the Sino-German Science Promotion Center (GZ295); Project 111 of State Administration of Foreign Experts Affairs and Ministry of Education, China (B06008); the Key Lab of Evolution of Past Life and Environment in NE Asia, Ministry of Education, China (Jilin Univ.); the Key Lab of Evolution of Past Life in NE Asia, Ministry of Natural Resources, China (Shenyang Normal Univ.); the NIGPAS, IVPP, and PMOL of China; the Office 305 of Xinjiang; the Geological Bureau and Survey of Xinjiang and RGSX No.1; the local governments of Turpan Region of Xinjiang, including the Shanshan County. Without their kind support and assistance, the authors' work could not have been carried out so successfully.

Sincere thanks also go to Academician Prof. Liu J. Q. (IGG CAS) and Prof. Wang Y. D. (NIGPAS) for their kind recommendation of this book for publication; to Academician Prof. Li T. D. (CAGS) for his warm support for this cooperative work; to Prof. Shen Y. C. (IGG CAS) and Dr. Cheng X. S. (CUG) for their great help to the authors' early work for the project; to Prof. Pospelov I. (IG RAS) for his kind help in finding the materials about Prof. Obruchev V. A.'s investigation trip to China in the VCEGEI, Russia.

Finally, the authors would cordially thank our good friends in Xinjiang, China and all members of the Sino-German Cooperative Research Team in Xinjiang, Who made their contributions to this project during the past more than twenty years from 1997 to 2020, including official leaders and scientists: Han J. G. , Chen L. S. and Lu R. K. (Beijing); Tian J. R., Wang B. L., Dong L. H., Zuo X. Y., Song S. S., Zhen B. S., Zhou H. Z., Li J., Du Q., Zhao W. Q., Liu K. Q., and Li Y. L. (Xinjiang); Dong Z. M., Xu X., Sha J. G., Cai C. Y., Wang C. Y., Shang Y. K., Wang W. M., Wang Y., Wang X. F., and Li S. J. (CAS); Sun C. L., Sun F. Y., Lv J. S., Chen Y. J., Li C. T., Wang P. J., Liu Z. J., Bian W. H., Liu R., Meng Q. T., Zhang J. G., Li L., and Sun W. (JU); Zhou C. F., and Zhang Y. (PMOL); Ma Y. (CNA); and Ye J. (Shanghai); German experts Pfretzschner U.-H., Bruch A., Eder J., Hinderer M., Hornung J., Thein J., Puettman W., Eitel B.,

Ligouis B., Wings O., Joyce W., Maisch M., Matzke A., Uhl D., Rabi M., Simone K., Regent V., and Mosbrugger G., Japanese young expert Sekiya T., etc. (Fig. 64). Many thanks would be extended to the Editor-in-Chief Wang S. P., Editor Wu H. L., and Art Designer Tang S. L. and Li M. X., from Shanghai Scientific & Technological Education Publishing House for their kind support and assistance for this book to be published successfully.

# References

Ashraf A R, Sun G, Wang X F, et al. 1998. Development of forest-ecosystems and palaeoclimate across the Triassic-Jurassic boundary in the Junggar Basin (NW China)—preliminary results. -The 5th EPPC, June 26-30, 1998 Carcow, Poland ; Wladyslaw Szafer Institute of Botany, Polish Acad Sci, 4.

Ashraf A R, Wang X F, Sun G, et al. 2001. Palynostratigraphic analysis of the Huangshanjie-, Haojiagou-and Badaowan formations in the Junggar Basin (NW China). *Proc. Sino-German Co-Symp. Prehist. Lie Geol. Junggar Bas. Xinjiang, China*, 40-64.

Ashraf A R, Sun Y W, Sun G. 2004. Guide booklet. *In*: Sun G, et al (editor-in-chief). Proceedings of Sino-German Cooperation Symposium on Paleontology, Geological Evolution, and Environmental Changes in Xinjiang, China. Urumqi. Special Material, 1-8.

Ashraf A R, Sun Y W, Sun G, et al. 2010. Triassic and Jurassic palaeoclimate development in the Junggar Basin, Xinjiang, Northwest China—a review and additional lithological data. *Palaeobio Palaeoenv*, 90: 187-201.

Bureau of Geology and Mineral Resources of Xinjiang (BGMRX). 1993. Regional geology of Xinjiang Uygur Zizhiqu. Beijing: Geological Publishing House, 1-841.

Cao R L, Zhu S H, Zhu X K, et al. 1993. Plate and terrain tectonics of northern Xinjiang. *In*: Tu G Z, et al (eds). New improvement of solid geosciences in northern Xinjaing. Beijing: Sci Press, 11-26.

Chen P J, McKenzie K G, Zhou H Z. 1996. A further research into Late Triassic *Kazacharthra* Fauna from Xinjiang Uygur Autonomous Region, NW China. *Act Palaeont Sin*, 35(3): 272-302.

Cheng Z W, Wu S Z, Fang X S, 1996. The Permian-Triassic sequences in the southern margin of the Junggar Basin and the Turpan Basin, Xinjiang, China. Beijing: Geological Publishing House, 1-25.

Deng S H, Cheng X S, Qi X F, et al. 2001. Late Triassic-early Jurassic plant assemblage in Junggar Basin, Xinjiang. *In*: Cheng Y Q (editor-in-chief). Proceedings of the 3rd National Stratigraphic Conference. Beijing: Geological Publishing House, 174-178.

Deng S H, Lu Y Z, Fan R, et al. 2010. The Jurassic System of Northern Xinjiang, China. Hefei: Univ Sci Techn China Press, 1-279.

Deng S H, Wang S E, Yang Z Y, et al. 2015. Comprehensive study of the Middle-Upper Jurassic strata in the Junggar Basin, Xinjiang. *Acta Geosci Sin*, 36(5): 559-574.

Doludenko M P, Russkazova E S. 1972. 19712 Ginkgoales and Czekanowskiales of the Irkutsk Basin. *In*: Mesozoic plants (Ginkgoales and Czekanowskiales) of East Siberia. Nauka, Moscow, 7-43. (in Russian)

Dong M. 2012 Middle Jurassic plants from Shaerhu Coal Field of Xinjiang, China. Ph. D. dissertation. Changchun: College of Earth Science of Jilin University, 1-98.

Dong M, Sun G. 2011. Middle Jurassic plants from Shaerhu coal field of Xinjiang, China. *Global Geology*, 30(4): 497-507.

Dong Z M. 1973. Dinosaurs from Wuerho. *Mem IVPP*, (11): 45-52.

Dong Z M. 1977. On the dinosaurian remains from Turpan, Xinjiang. *Vertebrata Palasiatica*. 15(1): 59-66.

Dong Z M. 1989. On a small Ornithopod (*Gongbusaurus wucaiwanensis* sp. nov.) from Karamaili, Junggar, Xinjiang, China. *Vertebr PalAsiatica*, 27(2):140-146.

Dong Z M. 1990. On remains of Sauropods from Karamaili region, Junggar, Xinjiang, China. *Vertebr PalAsiatica*, 28(1): 43-58.

Dong Z M. 1993. An ankylosaur (Ornithischian dinosaur) from Middle Jurassic of the Junggar, Basin, China. *Vertebr PalAsiatica*, 31(4): 257-266.

Dong Z M. 1997. A gigantic sauropod (*Hudiesaurus sinojapanorum* gen. et sp. nov.) from the Turpan Basin, China. *In*: Dong Z M(ed). *Sino-Japanese Silk Road Dinosaur Expedition*. Beijing: China Ocean Press, 102-110.

Dong Z M. 2001. Mesozoic fossil vertebrates from the Junggar Basin and Turpan Basin, Xinjiang, China. *Proc. Sino-German Co-Symp. Prehist. Lif. Geol. Junggar Bas. Xinjiang, China*: 95-103.

Dong Z M. 2009. Dinosaur in Asia. Kunming: Yunnan Science & Technology Press, 1-287.

Florin R. 1936. Die fossilen Ginkgophyten von Franz-Joseph-Land nebst Erorterungen uber vermeintliche Cordaitales mesozoischen Alters. I. *Palaeontographica* B, 81: 71-173.

Geological Survey of Xinjiang, Bureau of Geology and Mineral Resources of Xinjiang. 1983. Regional geological report (Baiyanghe area; 1:200000 in scale). Beijing: Geological Publishing House, 1-256.

Gu D Y. 1980. Fossil Plants. Paleontological Atlas of Xinjiang. Beijing: Geological Publishing House.

Halle A. 1908. Zur Kenntnis der mesozoischen Equisetales Schwedens. *K. Svensk. Vet. Akad. Handl.*, 7: 113.

Harris T M. 1951.The fructification of *Czekanowskia* and its allies. *Philophical Transactions of the Royal Society of London Series*, 235: 483-508.

Harris T M. 1961. The Yorkshire Jurassic Flora, I . *Thallophyte-Pteridophyta*. London: British Museum (Nat Hist), 1-204.

Harris T M, Millington W. 1974. The Yorkshire Jurassic flora. IV . 1. Ginkgoales; 2.

Czekanowskiales. London: British Museum of London, 2-150.

Hornung J, Sun G, Li J, et al. 2003. Fluvial-deltaic lithofacies and architectural element analyses at the Haojiagou-valley section in the Junggar-Basin (NW-China, Middle/Upper Triassic). *Palaeontographica*, in review.

Hsu R, et al. 1979. *Late Triassic flora of Baoding, China*. Beijing: Sci Press, 1-130.

Hu A Q, Zhang G X, Li Q X, et al. 1997. Isotopie geochemistry and crustal evolution of northern Xinjiang. *In*: Tu G Z, et al. (eds). *New improvement of solid geosciences in northern Xinjaing*. Beijing: Sci Press, 27-37.

Hu Y M, Meng J, Clark J M. 2007. A new Late Jurassic docodont (Mammalia) from northeastern Xinjiang, China. *Vertebr PalAsiatica*, 45: 173-194.

Huang B. 2006. Sporopollen assemblages from the Haojiagou and Badaowan formations at the Haojiagou section of Urumqi, Xinjiang and their stratigraphical significance. *Acta Micropalaeontologica Sinica*, 23(3): 235-274.

Huang Z G, Zhou H Q. 1980. Plants. *In: Stratigraphy and palaeontology of the Shaanxi, Gansu and Ningxia Basin*. Beijing: Geological Pulishing House, 185. (in Chinese)

Kang Y Z. 2003. Structural features of Junggar Tarim and Turpan-Hami Basins Xinjiang and the distribution of oil and gas. *Journal of Geomechanics*, 9(1): 37-47.

Kiritchkova A I, Travina T A, Bystritskaya L I. 2002. The *Phoenicopsis* Genus: Systematics. History, Distribution and Stratigraphic Significance. Petersburg: VNIGRI, 1-205 (in Russian)

Krassilov V A. 1968. On the study of fossil plants of Czekanowskiales. *Transactions of Geological Institute, Academy of Science USSR*, 191: 31-40. (in Russian)

Krassilov V A. 1972. Morphology and systematics of Ginkgoles and Czekanowskiales. *Paleontological Journal*, 1: 113-118. (in Russian)

Li C Y, Wang Q, Liu X Y, et al. 1982. *Explanatory notes to the tectonic map of Asia*. Beijing: Cartographic Press.

Li J, Zhen B S, Sun G. 1991. First discovery of Late Triassic florule in Wusitentag-Karamiran area of Kunlun Mountain of Xinjiang. *Xinjiang Geol*, 9(1): 50-58, 98-99.

Li P J, He Y L, Wu X W, et al. 1988. *Early and Middle Jurassic strata and their floras from northeastern boundary of Qaidam Basin, Qinghai*. Nanjing: Najing Univ Press, 1-231.

Li Y A, Jin X C, Sun D J, et al. 2003. Paleomagnetic Properties of Non-marine Permo-Triassic Transitional Succession of the Dalongkou Section, Jimsar, Xinjiang. *Geological Review*, 49(5): 525-536.

Ligouis B. 2001. Organic petrography, geochemistry, stratigraphy, facies geology and basin analysis in the Triassic and Jurassic of the Junggar-Basin (NW China). *Proc. Sino-German Co-Symp. Prehist. Lif. Geol. Junggar Bas. Xinjiang, China*, 104-113.

Maisch M W, Matzke A T, Pfretzschner H-U, et al. 2001. The fossil vertebrate faunas of the Toutunhe and Qigu formations of the southern Junggar Basin and their biostratigraphical and paleoecological implications. *Proc. Sino-German Co-Symp. Prehist. Lif. Geol. Junggar Bas. Xinjiang, China*, 83-94.

Maisch M W, Matzke A T, Sun G. 2004. A relict trematosauroid (Amphibia: Temnospondyli) from the Middle Jurassic of the Junggar Basin (NW China). *Naturwissenschaften*, 91 (12): 589–593.

Maisch M, Matzke A, Grossmann F, et al. 2005. The first haramiyid mammal from Asia. *Naturwissenschaften*, 92: 40-44.

Martin T. 2004. Incisor enamel microstructure of South America's earliest rodents: implications for caviomorph origin and diversification. *In*: The Paleogene Mammalian Fauna of Santa Rosa, Amazonian Peru. K. E. Campbell, Jr., editor, Natural History Museum of Los Angeles County, *Science Series*, 40: 131-140.

Martin T, Averianov A O, Pfretzschner H-U. 2010a. Mammals from the Late Jurassic Qigu Formation in the Southern Junggar Basin, Xinjiang, Northwest China. *Palaeodiv Palaeoenv*, 90: 295-317.

Martin T, Sun G, Mosbrugger V. 2010b. Triassic-Jurassic biodiversity, ecosystem, and climate in the Junggar Basin, Xinjiang, Northwest China. *Palaeobio Palaeoenv*, 90: 171-173.

Matzke A T, Maisch M W, Sun G, et al. 2004. A new Xinjiangchelyid turtle (Testudines, Eucryptodira) from the Jurassic Qigu Formation of the southern Jungar Basin, Xinjiang, North-West China. *Palaeontology*, 47:1267-1299.

Matzke A T, Maisch M W, Sun G, et al. 2005. A new Middle Jurassic xinjiangchelyid turtle (Testudines; Eucryptodira) from China (Xinjiang, Junggar Basin. *J Vertebr Paleont*, 25: 63-70.

Miao Y Y. 2003. Discovery of *Leptostrobus laxiflora* Heer from Middle Jurassic Xishanyao Formation in the Baiyang River of Emin, Xinjiang. *Jilin Univ (Earth Sci Edit)*, 33 (3): 263-269.

Miao Y Y. 2005. New material of Middle Jurassic plants from Baiyang River of northwestern Junggar Basin, China. *Act Palaeont Sin*, 44(4): 517-534.

Miao Y Y. 2006. Ginkgoales and Czekanowskiales from the Middle Jurassic in western Junggar Basin of Xinjiang. Changchun: College of Earth Science of Jilin University, 1-146.

Miao Y Y. 2017. Discussion on leaf cuticular characters of *Ginkgo obrutschewii* Seward from Middle Jurassic of Junggar Basin, Xinjiang, China. *Global Geology*, 36(1):15-21.

Nebelsick J. 2004. Carbonatic facies dynamics during period of global climatic transition. *Proc. Sino-Germ. Symp. Paleont. Geol. Evul. Env. Chang.*, Urumqi, 73-75.

Nosova N, Zhang J W, Li C S. 2010. Revision of *Ginkgoites obrutschewii* (Seward) Seward (Ginkgoales) and the new material from the Jurassic of Northwestern China. *Review of Palaeobotany and Palynology*, 166: 286-294.

Obruchev V A. 1914. Diam River, Orkhu Basin and Mt. Khara-Arati. Border area of Dzunggaria. Route Investigation. *Bulletin of Tomsk Technological Institute, Russia*, 1

(2):433-437.

Obruchev V A. 1940a. Geological map of boundary region of Dzungaria (1:500000 in scale). *In*: Obruchev V A. 1940. Boundary region of Dzungaria. M-L: Acad Sci, USSR, 3.

Obruchev V A. 1940b. Regional Dzunggaria. Geology. Geographical and Geological Records. Geological Science Institute, Acad Sci USSR. M-L:Acad Sci USSR, 3 (2): 138-156.

Ouyang H, Ye Y. 2002. *The first Mamenchisaurian skeleton with complete skull*: Mamenchisaurus youngi. Chengdu: Sichuan Science and Technology Press, 1-111.

Pfretzschner H-U, Ashraf A R, Maisch M, et al. 2001. Cyclic growth in dinosaur bones from the Upper Jurassic of NW China and its paleoclimatic implications. *Proc. Sino-German Co-Symp. Prehist. Lif. Geol. Junggar Bas. Xinjiang, China*, 21-39.

Pfretzschner H-U, Martin T, Maisch M, et al. 2004. A new docodent from the Totunhe Formation (Middle Jurassic) of the Junggar Basin. *Proc. Sino-Germ. Symp. Paleont. Geol. Evul. Env. Chang.*, Urumqi: 9-11.

Pfretzschner H-U, Martin T, Maisch M, et al. 2005. A new docodont mammal from the Late Jurassic of the Junggar Basin in Northwest China. *Acta Palaeontol Pol*, 50(4): 799-808.

Rabi M, Zhou C F, Wings O, et al. 2013. A new xinjiangchelyid turtle from the Middle Jurassic of Xinjiang, China and the evolution of the basipterygoid process in Mesozoic turtles. *BMC Evolutionary Biology*, 13: 203.

Regional Geological Survey of Xinjiang (RGSX), Geological Institute of Xinjiang, Geological Surveying Division of Petroleum Bureau of Xinjiang. 1983. Palaeontological Atlas of Northwest China. Vol. Xinjiang. Beijing: Geological Publishing House, 1-795.

Regional Geological Survey of Xinjiang. 1965. Geology of Urumqi areas (K-45-IV). Beijing Print Fact, 8: 1-35.

Samylina V A. 1967. On the final stages of the history of the genus *Ginkgo* L. in Eurasia. Z. Bot. 52: 303-316.

Samylina V A, Kiritchkova A I. 1991. The genus *Czekanowskia* (Systematics, History, Distribution and Stratigraphic Significance). Leningrad: Nauka. 1-130. (in Russian)

Samylina V A, Kiritchkova A I. 1993. The genus *Czekanowskia* Heer. Principles of systimatics, range in space and time. *Review of Palaeobotany and Palynology*, 79: 271-284.

Seward A C. 1911. Jurassic plants from Chinese Dzungaria, collected by Professor Obruchev. *Mem. Com. Geol. N.S. St. Petersbourg*, 75: 1-61.

Sun A L. 1978. Two new genera of Dicynodontidae. In Permian and Triassic vertebrate fossils of Dzungaria Basin and Tertiary stratigraphy and mammalian fossils of Turfan basin. *Mem Inst Vertebr Paleont Paleoanthropol, Acad Sin*, 13:19-25.

Sun C L, Dilcher D L, Wang H S, et al. 2009. *Czekanowskia* from the Jurassic of Inner Mongolia, China. *International Journal of Plant Sciences*, 170 (9): 1183-1194.

Sun C L, Na Y L, Dilcher D L, et al. 2015. A New Species of *Phoenicopsis* Subgenus

*Windwardia* (Florin) Samylina (Czekanowskiales) from the Middle Jurassic of Inner Mongolia, China. *Acta Geologica Sinica*, 89(1): 55-69.

Sun G. 1979. On the discovery of *Cycadocarpidium* from the Upper Triassic of eastern Jilin. *Act Palaeont Sin*, 18 (3): 312-326.

Sun G. 1987. On Late Triassic geofloras in China and principles for palaeophytogeographic regionalization. *Act Geol Sin*, 61(1):1-9.

Sun G. 1993a. *Ginkgo coriacea* Florin from Lower Cretaceous of Huolinhe, northeastern Inner Mongolia, China. *Palaeontographica B*: 159-168.

Sun G. 1993b. *Late Triassic flora from Tianqiaoling of Jilin, China*. Changchun: Jilin Sci Techn Press, 1-157.

Sun G, Meng F S, Qian L J, et al. 1995. Triassic floras of China. *In*: Li X X (eds-in-chief). *Fossil floras of China through the geological ages*. Guangzhou: Guangdong Sci Techn Press, 229-259.

Sun G, Mosbrugger V, Ashraf A R, et al. 2001a. The advanced study of prehistory life and geology of Junggar Basin, Xinjiang, China. Urumqi, 1-113.

Sun G, Mosbrugger V, Li J, et al. 2001b. Late Triassic flora from the Junggar Basin, Xinjiang, China. *Proc. Sino-German Co-Symp. Prehist. Lif. Geol. Junggar Bas. Xinjiang, China*, 8-20.

Sun G, Miao Y Y, Sun Y W, et al. 2004a. Middle Jurassic plants from Baiyang River area of Enin, northwesterm Junggar Basin, Xinjiang. *In*: Sun G, et al (eds). Proc. Sino-German Symp. Geol. Evol. Environm. Changes of Xinjiang, China. 2004: 35-40.

Sun G, Mosbrugger V, Ashraf A R, et al. 2004b. Paleontology, geological evolution and environmental changes of Xinjiang, China. *Proc. Sino-Germ. Symp. Paleont. Geol. Evol. Env. Chang., Urumqi*: 1-96.

Sun G, Miao Y Y, Chen Y J. 2006. A new species of *Sphenobaiera* from Middle Jurassic of Junggar Basin, Xinjiang, China. *Jilin Univ (Earth Sci Edit)*, 36 (5): 717-722.

Sun G, Miao Y Y, Mosbrugger V, et al. 2010. The Upper Triassic to Middle Jurassic strata and floras of the Junggar Basin, Xinjiang, Northwest China. *Palaeobio Palaeoenv*, 90: 203-214.

Sun G, Yang T, Tan X, et al. 2021. Recent advances in the study of the ferns from Middle Jurassic Baiyanghe flora of Junggar Basin, Xinjiang, China. *Palaeobio Palaeoenv*, 101: 19-24.

Sze H C. 1933. Fossile Pflanzen aus Shensi, Szechuan und Kuuichow. *Palaeont Sin*, Ser A, 1 (3): 1-32.

Sze H C, Li X X. 1963. Fossil Plants of China (2). Mesozoic plants of China. Beijing: Sci Press, 1-429.

Tang T F, Yang H R, Lan X, et al. 1989. Cretaceous-Eogene marine strata and oil bearing properties in Western Tarim Basin, Xinjiang. Beijing: Sci Press.

Taylor T N, Taylor E L. 2009. The Biology and Evolution of Fossil Plants. Englewood Cliffs: Prentice Hall, 744-787.

Vakhrameev V A. 1988. Jurassic and Cretaceous floras and climates of the Earth. Nauka: 1-210. (in Russian)

Vakhrameev V A, Doludenko M P. 1961. Late Jurassic and early Cretaceous flora from Bureja Basin and its significance for stratigraphy. *Trans Geol Inst Acad Sci USSR*, 54: 1-135.

Wang S E, Gao L Z. 2012. HRIMP U-Pb dating of zircons from tuff of Jurassic Qigu Formation in Junggar Basin, Xinjiang. *Geological Bulletin of China*, 31(4): 503-509.

Wang T, et al. 1996. *Oil and gas reservoir geology of rift basins in China*. Beijing: Petroleum Industry Press, 1-172.

Wang X L, Kellner A W, Jiang S, et al. 2014. Sexually dimorphic tridimensionally preserved pterosaurs and their eggs from China. *Current Biology*, 24 (12): 1323-1330.

Wang X L, Kellner A W, Cheng X, et al. 2017. Egg accumulation with 3D embryos provides insight into the life history of a pterosaur. *Science*, 358(6367): 1197-1201.

Wang Y D, Zhang W, Saiki K. 2000. Fossil wood from the Upper Jurassic of Qitai, Junggar Basin, Xinjiang, China. *Acta Paleontologica Sinica*, 39 (Suppl.): 176-185.

Wang Y D, Shi Y, Miao Y Y. 2002. New advances of palaeontology and geology in the Junggar Basin, Xinjiang. *Episodes*, 25 (2): 128-131.

Wang Z, Zhan J Z, Li M. 1999. Cretaceous stratigraphic division in Eastern Tarim Basin. *Xinjiang Petroleum Geology*, 20(3).

Wings O, Schellhorn R, Mallison H, et al. 2007. The first dinosaur tracksite from Xinjiang, NW China (Middle Jurassic Sanjianfang Formation, Turpan Basin ) —a preliminary report. *Global Geology*, 10(2): 113-129.

Wings O, Rabi M, Schneider J W, et al. 2012. An enormous Jurassic turtle bone bed from the Turpan Basin of Xinjiang, China. *Naturwissenschaften*, 99 (11): 925-935.

Wu C Y, Xue S H, et al. 1997. Sedimentology of petroliferous basins in China. Beijing: Petroleum Industry Press, 401-411.

Wu S Q, Zhou H Z. 1986. Early Jurassic plants from East Tianshan Mountain. *Act Palaeont Sin*, 25 (6): 636-647.

Wu S Z, Li Y A, Qu X, et al. 2001. Triassic geography and climate of the Junggar Basin. *In*: Sun G, et al (eds). Proc. Sino-German Symp. Prehist. Lif. Geol. Junggar Basin, Xinjiang, China, 1-9.

Wu S Z. 1990. Characteristics of Early Jurassic flora in Xinjiang. *Xinjiang Geol*, 8 (2): 119-132.

Wu S Z, Liu Z Y. 1988. Cretaceous strata and sedimentary features in northern Xinjiang. *Xinjiang Geology*, 6(1): 21-30.

Wu W H, Zhou C F, Wings O, et al. 2013. A new gigantic sauropod dinosaur from the Middle Jurassic of Shanshan, Xinjiang. *Global Geology*, 32(3): 438-462.

Xi D P, Wan X Q, Li G B, et al. 2019. Cretaceous integrative stratigraphy and timescale of China. *Journal of Earth Science China*, 62 (1), 256-286.

Xi D P, Tang Z H, Wang X J, et al. 2020. The Cretaceous Paleogene marine stratigraphic framework that records significant geological events in the western Tarim Basin. *Earth Science Frontiers*, 27(6): 165-198.

Xing L D, Klein H, Lockley M G, et al. 2014. *Changpeipus* (theropod) tracks from the Middle Jurassic of the Turpan Basin, Xinjiang, Northwest China: review, new discoveries, ichnotaxonomy, preservation and paleoecology. *Vertebrata PalAsiatica*, 52(2): 233-259.

Xu X, Clark J M, Forster C A, et al. 2006. A basal tyrannosauroid dinosaur from the Late Jurassic of China. *Nature*, 439: 715-718.

Yang J D, Qu L F, Zhou H Q, et al. 1986. Permian and Triassic strata and fossil assemblages in the Dalongkou area of Jimsar, Xinjiang. *Geol. Mem.* Ser. 2, No. 3, Beijing: Geological Publishing House, 1-262.

Yang J L, Shen Y X. 2004. The spatial and temporal distribution and cause explanation of the Ziniquanzi Formation from Junggar Basin. *J Stratigraphy*. 28: 215-222.

Yang J L, Wang Q F, Lu H N. 2008. Cretaceous Charophyte floras from the Junggar Basin, Xinjiang, China. *Acta Micropaleontologica Sinica*, 25(4): 345-363.

Yang J L, Shen Y X, Shang H. 2012. Cretaceous-Palaeogene ostrocods from the southern edge of the Junggar Basin and their stratigraphic significance. *Acta Palaeontologica Sinica*, 51(3): 359-369.

Yang T, Xing M D, Bai S C. 2017a. Discovery of Czekanowskia baikalica from Middle Jurassic of Baiyanghe area, Xinjiang, China. *Shenyang Normal University (Natural Science Edition)*, 35(4): 288-302.

Yang T, Yi L Y, E J W, et al. 2017b. A new species of Scarburgia from Middle Jurassic of Junggar Basin, Xinjiang, China. *Global Geology*, 36(2): 327-332.

Yang T, Zhang Y J, E J W. 2018. Study of the epigenetic structure of Ferganiella from middle Jurassic Xishanyao Formation in Junggar Basin, Xinjiang. *Geology and Resources*, 27(2): 124-129.

Young C C. 1935. On two skeletons of Dicynodontia from Sinkiang. *Bull Geol Soc China*, 14: 483-517.

Young C C. 1978. A Late Triassic vertebrate fauna from Fukang, Sinkiang. *Mem Inst Vertebr Paleont Paleoanthropol, Acad Sin*, 13: 60-67.

Yuan P L, Young C C. 1934. On the discovery of new *Dicynodon* from Sinkiang. *Bull Geol Soc China*, 13: 563-574.

Zhang C, Yu X H, Yao Z Q, et al. 2021. Sedimentary evolution and controlling factors of the Middle-Upper Jurassic in the western part of the southern Junggar Basin. *Geology in China*, 48(1): 284-296.

Zhang X Q, Li D-Q, Xie Y, et al. 2018. Redescription of the cervical vertebrae of the mamenchisaurid sauropod *Xinjiangtitan shanshanesis* Wu et al. 2013. *Historical Biology*, 32 (5): 1-20.

Zhao X J. 1980. Mesozoic vertebrate fossil strata in Northern Xinjiang. *Bull IVPP A*, 15: 1-120.

Zhao X J. 1980. Mesozoic vertebrate fossils and strata from the northern Xinjiang. *Mem Inst Vertebr Paleont Paleoanthropol, Acad Sin*, 15: 1-120.

Zhen B S, Li J, Sun G. 1991. First discovery of the Late Triassic from Wustentag-Haramilan of Kunlun Mt., Xinjiang. Geol. Xinjiang, 9: 1-6.

Zheng X L, Zheng X M, Zheng X S, et al. 2013. The Late Cretaceous Ostrocod fossils of the Junggar Basin. *J Stratigraphy*, 37(2): 206-209.

Zhou C F, Bhullar B A S, Neander A I, et al. 2019. New Jurassic mammaliaform sheds light on early evolution of mammal-like hyoid bones. *Science*, 365: 276-279.

Zhou Z Y, Dean W T. 1996. Phanerozoic geology of Northwest China. Beijing: Sci Press, 1-316.

Zhou Z Y, Lin H L. 1995. Stratigraphy, paleogeography and plate tectonics from Northwest China. Nanjing: Nanjing Univ Press, 1-299. (in Chinese)

Zhou Z Y. 1995. Jurassic flora. *In*: Li X X (eds-in-chief). *Fossil floras of China through the geological ages*. Guangzhou: Guangdong Sci Techn Press, 343-410.

Zhou Z Y. 2001. Strata of Tarim Basin. Beijing: Sci Press, 1-359.

Zhou Z, Zhan B L. 1989. A Middle Jurassic Ginkgo with ovule-bearing organs from Henan, China. *Palaeontographica Abt* B, 211: 113-133.

# 主要英文缩写

# Main abbreviations in English

SGWSGX: Sino-German Co-Working Station for Geosciences in Xinjiang
中德合作新疆地质工作站

RGSX No. 1: Regional Geological Survey of Xinjiang No.1, China
新疆地矿局第一区调大队

RGSIX: Geological Surveying Institute of Xinjiang, China 新疆地调院

NSFC: Natural Science Foundation of China 中国国家自然科学基金委员会

SGSPC: Sino-German Science Promoting Center 中德科学中心

CAS: Chinese Academy of Sciences 中国科学院

NIGPAS: Nanjing Institute of Geology & Paleontology, CAS
中国科学院南京地质古生物研究所

IVPP: Institute of Vertebrate Paleontology & Paleoanthropology, CAS
中国科学院古脊椎动物与古人类研究所

PSC: Paleontological Society of China 中国古生物学会

PMOL: Paleontological Museum of Liaoning, China 辽宁古生物博物馆

JU: Jilin University of China 吉林大学

RCPS: Research Center of Paleontology & Stratigraphy, Jilin University of China
吉林大学古生物学与地层学研究中心

SYNU: Shenyang Normal University 沈阳师范大学

DFG: Science Foundation of Germany 德国科学基金会

SGN: Senckenberg Gesellschaft fuer Naturforschung, Germany 森肯堡自然科学协会

TU: University of Tuebingen, Germany 德国图宾根大学

BU: University of Bonn, Germany 德国波恩大学

IGMP: Institute of Geology, Minerology & Paleontology, University of Bonn
德国波恩大学地质、矿产与古生物研究所

# 作者简介

# Brief Introductions to the Authors

**孙革** 古植物学家，沈阳师范大学教授、博士，自然资源部东北亚古生物演化重点实验室主任，中国古生物学会副监事长。曾任中科院南京古生物所副所长，吉林大学古生物研究中心主任，辽宁古生物博物馆馆长及沈阳师范大学古生物学院院长，中德合作新疆地质工作站中方主任，中德古生物与地质联合实验室中方主任。李四光地质科学奖获得者（2005）。主要研究领域：古植物学与事件地层学。

**Prof. Dr. Sun G**. Shenyang Normal University, China. Director of Key–Lab of Evolution of Past Life in NE Asia, Ministry of Natural Resources, China. He was Deputy Director of NIGPAS, Co–Director of the SGWSGX, Co–Director of Sino–German Co–Lab for Geosci. Paleont., Director of RCPS （JU）, Director of PMOL and College of Paleont., SYNU. Receiving Award of Geosciences named Li S. G., China in 2005. Major: Paleobotany and Event Stratigraphy.

**莫斯布鲁格**（Mosbrugger V.）古植物学及古气候学家，博士，德国科学院院士，法兰克福大学教授，中国吉林大学及沈阳师范大学名誉教授；前森肯堡自然科学博物馆总馆长 (2005~2020)；曾任图宾根大学地质古生物学院院长、图宾根大学副校长、中德合作新疆地质工作站德方主任、中德古生物与地质联合实验室德方主任。德国"莱伯尼茨奖"获得者（2004）。主要研究领域：古植物学与古气候学。

**Prof. Dr. Mosbrugger V.** German Academician, Senckenberg Natural Institute, Germany. Ex-Director General of Senckenberg Museum（NH）（2005–2020）; Adjunct Prof. of Univ. Frankfurt, Germany; Honorary Prof. Jilin Univ., and Honorary Prof. Shenyang Normal Univ. China; Vice-President of Univ. Tuebingen; Co-Director of the SGWSGX, China; Co-Director of Sino-German Co-Lab for Paleont. Geosci. Receiving Leibnitz Award, Germany（2004）. Major: Paleobotany and Paleoclimatology.

**孙跃武** 古植物学与地层学专家，教授，博士，吉林大学古生物研究中心主任，中国古生物学会常务理事兼古植物学分会副理事长，东北亚国际地学研究与教学中心常务副主任。曾任中国古生物学会副秘书长、中德合作新疆地质工作站副主任等。在组织中德合作新疆野外工作中作出重要贡献。主要研究领域：古植物学、孢粉学及地层学。

**Prof. Dr. Sun Y. W.** Jilin University, China. Director of Research Center of Paleont. & Stratgr., Jilin Univ.（RCPS）since 2018; Executive Director of International Center of Geosci. Res. Educ., NE Asia, Vice Co-Director of the SGWSGX, China. Making important contributions to organizing Sino-German research team in Xinjiang field work. Major: Paleobotany, Palynology and Stratigraphy.

阿什拉夫（Ashraf A. R.）孢粉学家，博士，德国波恩大学教授；吉林大学、沈阳师范大学名誉教授。曾任喀布尔大学校长，马尼拉大学名誉教授，中德合作新疆地质工作站副主任。在吉林大学、沈阳师范大学均设有"阿什拉夫实验室"；在中国黑龙江嘉荫设有他的塑像，以表彰他对 K-Pg 界线研究所作出的贡献。主要研究领域：孢粉学及地层学。

**Prof. Dr. Ashraf A. R.** Univ. Bonn, Germany. Honorary Prof. of Jilin Univ., and Shenyang Normal Univ., China; Vice Co-Director of the SGWSGX, China. Honorably granted with the "Ashraf Lab" in both JU and SYNU. A statue was erected in Jiayin of Heilongjiang, China in honor of his contributions to the study of the KPgB. Major: Palynology and Stratigraphy.

马丁（Martin T.）古脊椎动物学家，德国波恩大学教授，沈阳师范大学客座教授。波恩大学地质古生物所前所长，曾任中德合作新疆地质工作站副主任（2005~2013）。是新疆侏罗纪准噶尔兽和瘤棳齿兽以及迄今世界最早具完整舌骨的微小柱齿兽的主要发现者。主要研究领域：古脊椎动物学与地层学。

**Prof. Dr. Martin T.** Univ. Bonn, Germany. Ex-Director of Inst. Geol. Paleont., Univ. Bonn（2007-2011）: Vice Co-Director of the SGWSGX, China; Adjunct Prof. of Shenyang Normal Univ., China. Leading the research group to discover the early mammals *Dsungarodon* and *Sineleutherus* in Xinjiang（2005, 2010）, and co-describing the mammaliaform *Microdocodon* from Inner Mongolia, China（2019）with the earliest known completely preserved hyoid-bones. Major: Vertebrate Paleontology and Ecomorphology.

苗雨雁 古植物学青年专家，编审，博士，北京自然博物馆展览部主任，中国古生物学会科普工作委员会副秘书长，中国古生物学会古植物学分会理事。曾任北京自然博物馆《大自然》杂志编辑部主任。在新疆中侏罗世白杨河植物群研究中取得突出成绩。主要研究领域：古植物学及地层学。

**Prof. Dr. Miao Y. Y.** Beijing Natural History Museum, China. Head of Dept. Exhibition, BNHM; Director of Edition of journal *Nature China*（2010–2018）; Ex-Vice-Secretary of Science Popularization Division, PSC （since 2021）; Council Member of Paleobotanical Committee of PSC, China（since 2016）. Achieved in study of Middle Jurassic Baiyanghe flora of Xinjiang. Major: Paleobotany and Stratigraphy.

王克卓 新疆区域地质学专家，教授级高工。曾任新疆第一区调大队总工程师、新疆地调院院长，中德合作新疆地质工作站常务副主任。为中德新疆地质工作站的运行作出重要贡献。主要研究领域：区域地质学及构造地质学。

**Prof. Dr. Wang K. Z.** Regional Geological Surveying Institution of Xinjiang（RGSIX）, China. Ex-Chief Geologist of Regional Geological Survey No.1, Xinjiang; Executive Vice Director of the SGWSGX. Making great contributions to the SGWSGX working. Major: Regional Geology and Tectonic Geology.

**吴文昊** 古脊椎动物学青年专家，副教授，博士，吉林大学古生物研究中心副主任，中国古生物学会副秘书长。在"新疆巨龙"研究中取得突出成绩。主要研究领域：古脊椎动物学与地层学。

**Assoc. Prof. Dr. Wu W. H.** Jilin University, China. Vice-Director of RCPS（since 2018）；Vice-Secretary of PSC；Achieved in study of the dinosaurs including *Xinjiangotitan* in Xinjiang. Major: Vertebrate Paleontology and Stratigraphy.

**杨涛** 古植物学青年专家，副教授、博士，沈阳师范大学古生物学院地质教研室主任。中德合作新疆地质工作站接待德国专家主要联系人。自 2015 年起，在新疆中侏罗世白杨河植物群研究中取得突出成绩。主要研究领域：古植物学及地层学。

**Assoc. Prof. Dr. Yang T.** Shenyang Normal University, China. Head of Dept. Geology, College of Paleontology, SYNU. Making contributions to the reception work in the SGWSGX. Achieved in study of the Middle Jurassic Baiyanghe flora of Xinjiang since 2015. Major: Paleobotany and Stratigraphy.

**董曼** 古植物学青年专家，长江大学讲师。2010 年于吉林大学获博士学位（导师：孙革教授）。在新疆沙尔湖中侏罗世植物群研究中取得突出成绩。主要研究领域：古植物学及地层学。

**Dr. Dong M.** Lecturer of Changjiang University, China（in Wuhan）. Receiving MSc. and Ph. D. in Jilin Univ., China in 2010, supervised by Prof. Dr. Sun G. Achieved in study of the Middle Jurassic Shaerhu flora of Turpan-Hami Basin, Xingjiang. Major: Paleobotany and Stratigraphy.